28 50

METAL FATIGUE
IN ENGINEERING

METAL FATIGUE IN ENGINEERING

H. O. Fuchs
Emeritus Professor of Mechanical Engineering
Stanford University
Stanford, California

R. I. Stephens
Professor of the Materials Engineering Division
and of the Mechanical Engineering Program
The University of Iowa
Iowa City, Iowa

A Wiley-Interscience Publication

JOHN WILEY & SONS
New York • Chichester • Brisbane • Toronto

"We hope this book will be used in making decisions about the design and operation of machines and structures. We have tried to state the facts and our opinions correctly, clearly, and with their limitations. But because of the uncertainties inherent in the material and the possibility of errors we cannot assume any liability. We urge the readers to spend effort in tests and in verification commensurate with the risks that they will assume."

Library of Congress Cataloging in Publication Data

Fuchs, Henry Otten, 1907-
   Metal fatigue in engineering.

   "A Wiley-Interscience publication."
   Includes indexes.
   1. Metals—Fatigue.   I.   Stephens, Ralph Ivan,
joint author.   II.   Title.

TA460.F78   620.1'63   80-294
ISBN 0-471-05264-7

Printed in the United States of America

10 9 8 7 6 5 4 3 2 1

To our colleagues in the SAE Fatigue Design and Evaluation Committee and the ASTM Committees E-9 on Fatigue and E-24 on Fracture Testing, and to our families who suffered while we wrote.

PREFACE

This book is written for engineers and engineering students concerned with the design and development of equipment subjected to repeated loadings and for others who must make decisions concerning the fatigue resistance of a structure, machine, or component. It is intended for self-study by practicing engineers and as a textbook for a regular one semester or one quarter senior or graduate level course in mechanical, civil, or materials engineering, and for short courses on the subject. The material is applicable to many structures and machines, including automobiles, aerospace vehicles, bridges, tractors, ships, and nuclear pressure vessels.

The focus of this book is on applied engineering aspects of structures, machines, and components. There is enough background information to enable readers to judge the validity of the book's recommended procedures, but topics of interest principally to researchers, and not yet applied in hardware design, have been omitted. The methods presented have been used successfully in fatigue design.

The book integrates linear elastic fracture mechanics (LEFM) with notch strain analysis and S–N methodology. We recognize that neither concept alone adequately explains the fatigue design situation. At least two different fatigue criteria are needed, because fatigue failures proceed in at least two stages with different features: early development of cracks by localized plastic shear strain and later growth of cracks by more distributed tensile stresses. A third stage, final fracture, completes the overall fatigue process.

Fatigue design with metals is the primary subject of this book. The importance of self-stresses (residual stresses) and stress concentrations is emphasized. These concepts provide a key to successful fatigue design. Modern treatments of cumulative damage, or life prediction, that recognize the importance of cycle counting and sequence of events are included. To show that fatigue knowledge is a human endeavor rather than just a collection of laws of nature, footnotes giving biographical data about people such as Wöhler, Goodman, Griffith, Peterson, and Weibull are included. References are included in each chapter. Appendices on fatigue and fracture properties for selected engineering materials appear at the back of the book. Use is made of these appendices both in the text and in the problems. The book is written in SI units, with British/American units used where appropriate. Most figures, tables, and appendices have dual units. Illustrative problems are included to better show the applicability of the material to engineering design situations. Unsolved problems are included at the end of most chapters, and in many cases more than one answer or a range of answers is reasonable because of different modeling, methods of solution, material properties chosen, and the inherently nonexact nature of fatigue design. The problems often require the reader to indicate the significance of a solution and to comment on the assumptions made. Most chapters have a "Dos and Don'ts in Design" section.

Chapter 1 provides an historical overview of the subject. Chapter 2 discusses general fatigue design procedures, such as safe-life, fail-safe, forecasting of service reliability, monitoring, and inspection. Chapter 3 describes macroscopic and microscopic aspects of fatigue behavior, and Chapter 4 provides the needed fundamentals of LEFM principles to determine fatigue crack growth and final fracture. Chapter 5 gives constant amplitude fatigue data obtained from laboratory tests, such as stress-life (S–N), strain-life (ϵ–N), and crack propagation (da/dN–ΔK), for use in predicting fatigue behavior of machine parts and structures. Scatter of data and its treatment by statistical methods are examined. Chapter 6 discusses the importance and effect of stress concentrations in fatigue design, and Chapter 7 considers self-stresses (residual stresses) as well as notch strain analysis. Chapter 8 uses primarily the results of Chapters 5 to 7 to predict fatigue life and fatigue strengths of components under constant amplitude loading, a basis for succeeding chapters. Multiaxial fatigue is covered in Chapter 9. Fatigue from real load histories is the subject of Chapter 10, which treats cumulative damage, counting techniques, interaction models, and simulation testing. Environmental effects, including corrosion, fretting, temperature, and neutron irradiation, are discussed in Chapter 11. Chapters 12 and 13 involve fatigue of joints and components.

The fatigue resistance or fatigue life of metal parts and structures can be predicted by the methods and data in this book. The precision of the predictions will depend on the available supporting services, such as computer programs for finite element analyses and cycle counting. The accuracy of such predictions must be verified by tests and inspection because the actual service loads and environments, as well as the exact properties of the material used in production, may differ from those assumed in the computations.

We acknowledge our debt to many workers in the field by references to publications. Many people who are not mentioned, however, have contributed to the knoweldge of fatigue. Discussions with colleagues on university faculties and particularly the SAE and ASTM committees have helped to refine our understanding. Our families have endured the diversion of our energies to this book. Some financial support was provided to R. I. Stephens by an Old Gold Faculty Fellowship from the University of Iowa. For all this help, we are sincerely grateful. We shall also be grateful to those readers who inform us of any errors that they find in the book.

Stanford, California   H. O. FUCHS
Iowa City, Iowa   R. I. STEPHENS
April 1980

CONTENTS

METAL FATIGUE
IN ENGINEERING

CHAPTER 1

HISTORICAL OVERVIEW

Mechanical failures have caused many injuries and much financial loss. Failure due to repeated loading, that is fatigue, has accounted for at least half of these mechanical failures. No exact percentage is available, but many books and articles have suggested that between 50 and 90 percent of all mechanical failures are fatigue failures; most of these are unexpected fractures. This challenges the engineer to improve design decisions involving fatigue.

Many approaches to fatigue design exist. They can be simple, inexpensive approaches or they may be extremely complex, sophisticated, and expensive. A more complete analysis may initially be more expensive, but in the long run it may be the least expensive. Thus an important question in fatigue design is how complete the analysis should be. Current product liability laws have placed special emphasis on explicitly documented design decisions.

Fatigue of materials is still only partly understood. What we do know has been developed step by step and has become quite complex. To gain some general understanding one starts best with a brief historical review of fatigue developments. This shows a few basic ideas and indicates very briefly how they were developed by the efforts of many people.

The first major impact of failures due to repeated stresses involved the railway industry in the 1840s. It was recognized that railroad axles failed regularly at shoulders [1]. Even then, the elimination of sharp corners was recommended. Since these failures appeared to be quite different from normal ruptures associated with monotonic testing, the erroneous concept of "crystallization" due to vibration was suggested but was later refuted. The word "fatigue" was introduced in the 1840s and 1850s to describe failures occurring from repeated stresses. This word has continued as the normal description of fracture due to repeated stresses. In Germany during the 1850s and 1860s August Wöhler* performed many laboratory fatigue tests under repeated stresses. These experiments were concerned with railway axle failures and are considered to be the first systematic investigation of fatigue. Thus Wöhler has been called the "father" of systematic fatigue testing. He showed from stress versus life (S–N) diagrams how fatigue life decreased with higher stress amplitudes and that below a certain stress amplitude, the test specimens did not fracture. Thus Wöhler introduced the concept of the S–N diagram and the fatigue limit. He pointed out that for fatigue the range of stresses is more important than the maximum stress [2]. During the 1870s and 1890s additional researchers substantiated and expanded Wöhler's classical work. Gerber along with others, investigated the influence of mean stress, and Goodman† proposed a simplified theory concerning mean stresses. Their names are still associated with diagrams involving alternating and mean stresses.

In the 1900s the optical microscope was used to pursue the study of fatigue mechanisms. Localized slip lines and slip bands leading to the formation of microcracks were observed. In the 1920s Gough‡ and associates contributed heavily to the understanding of fatigue mechanisms. They also showed the combined effects of bending and torsion (multiaxial fatigue). Gough published a comprehensive book on fatigue of metals in 1924 [3]. Moore and Kommers [4] published the first comprehensive American book

---

* August Wöhler (1819–1914). After graduating from the Technical University of Hanover, Germany and working on railways, he became chief of rolling stock of the Berlin to Breslau railroad in 1847. From 1847 to 1889 he was in Strasbourg as director of Imperial Railroads. In 1870 he stated that the stress range is decisive for fatigue failures. His exhibit of fatigue test results at the Paris exhibition of 1867 was perceptively reviewed in *Engineering*, Vol. 2, 1867, p. 160.

† John Goodman (1862–1935). He was Professor of civil and mechanical engineering at the University of Leeds, England and published the widely used text *Mechanics Applied to Engineering*, 1st ed., 1904, 8th ed., 1914, in which he says ". . . it is assumed that the varying loads applied to test bars by Wöhler and others produce the same effects as suddenly applied loads. . . . the only excuse for adopting such a theory is that it gives results fairly in accord with experimental values and moreover it is easily remembered and applied." Some modern authors unfortunately use Goodman's assumption as if it were a law of nature.

‡ Herbert J. Gough (1890–1965). He received his engineering degrees from University College School and London University in England, including the D.Sc. and Ph.D. He joined the National Physical Laboratory in England in 1914 but then spent the next 5 years involved in World War I. Gough returned to the scientific staff at NPL and became superintendent of the engineering department from 1930 to 1938. During World War II he was director of scientific research in the War Department followed by an appointment in the Ministry of Supply. He was President of the Institute of Mechanical Engineers in 1949 and had more than 80 publications on fatigue of metals along with many international lectures and awards.

on fatigue of metals in 1927. In 1920 Griffith* [5] published the results of his theoretical calculations and experiments on brittle fracture using glass. He found the strength of glass depended on the size of microscopic cracks. If S is the nominal stress at fracture and a is the crack size at fracture the relation is $S\sqrt{a}$ = constant. By this classical pioneering work on the importance of cracks Griffith became the "father" of fracture mechanics.

In 1929/1930 Haigh† [6] presented his rational explanation of the difference in the response of high tensile strength steel and of mild steel to fatigue when notches are present. He used concepts of notch strain analysis and self-stresses that were later more fully developed by others. During the 1930s an important practical advance was achieved by the introduction of shot-peening in the automobile industry. Where fatigue failures of springs and axles had been common, they then became rare. Almen‡ [7] correctly explained the spectacular improvements by compressive stresses produced in the surface layers of peened parts and promoted the use of peening and other processes that produce beneficial self-stresses. Horger [8] showed that surface rolling can prevent the growth of cracks. In 1937 Neuber§ [9] introduced stress gradient effects at notches and the elementary block con-

---

* A. A. Griffith (1893–1963). He graduated from the University of Liverpool, England in 1921 with the degrees of B. Eng., M. Eng., and D. Eng. He entered the Royal Aircraft factory in 1915 and advanced through a workshop traineeship followed by other positions to become senior scientific officer in 1920. In 1917, together with G. I. Taylor, he published a pioneering paper on the use of soap films in solving torsion problems, and in 1920 he published his famous paper on the theory of brittle fracture. He then worked on the design theory of gas turbines. Griffith was Head of the Engine Department of the Royal Aircraft Establishment in 1938 and joined Rolls Royce as research engineer in 1939. He worked first on conceptual design of turbojet engines and later on vertical takeoff aircraft design. He retired in 1960 but continued working as a consultant for Rolls Royce.

† Bernard P. Haigh (1884–1941). Haigh was born in Edinburgh, Scotland and received engineering training at the University of Glasgow. He received the D.Sc. degree. During World War I he served with the Admiralty Service and beginning in 1921 he was professor of applied mechanics at the Royal Naval College. He produced many inventions and many scientific papers on engineering subjects including significant contributions to fatigue of metals.

‡ John Otto Almen (1886–1973). Born in a sod cabin in North Dakota, he graduated from Washington State College in 1911 and became a prolific inventor. He joined General Motors Research Laboratories in 1926. Almen said he turned to the study of residual stresses to escape from being second-guessed by administrators who thought they could improve his mechanical inventions. According to Almen, "fatigue failures are tensile failures."

§ Heinz Neuber (1906). He is a graduate of Technical University, Munich, Germany, professor of applied mechanics there, and director of its Mechanical Technology Laboratory.

cept, which considers that the average stress over a small volume at the root of the notch is more important than the peak stress at the notch.

During World War II the deliberate use of compressive self-stresses became common in the design of aircraft engines and armored vehicles. Many brittle fractures in welded tankers and in Liberty ships motivated substantial efforts and thinking concerning preexisting defects in the form of cracks and the influence of stress concentrations. Many of these brittle fractures started at square hatch corners or square cutouts and welds. Solutions included rounding and strengthening corners, adding riveted crack arresters, and greater emphasis on material properties. In 1945 Miner [10] formulated a linear cumulative fatigue damage criterion suggested by Palmgren [11] in 1924. This linear fatigue damage criterion is now recognized as the Palmgren–Miner law. It has been used extensively in fatigue design and, despite its many shortcomings, still remains an important tool in fatigue life predictions.

The Comet, the first jet propelled passenger airplane, started service in May 1952 after more than 300 hours of flight tests. Four days after an inspection in January 1954 it crashed into the Mediterranean Sea. After much of the wreckage had been recovered from the bottom of the sea and exhaustive investigation and tests on components of Comet aircraft had been made, it was concluded that the accident was caused by fatigue failure of the pressurized cabin. The small fatigue cracks originated from a corner of an opening in the fuselage. Two Comet aircraft failed catastrophically. The Comet had been tested thoroughly. The cabin pressure at high altitudes was 57 kPa (8.25 psi) above outside pressure. By September 1953 a test section of the cabin had been pressurized 18,000 times to 57 kPa in addition to 30 earlier tests at between 70 and 110 kPa. The design stress for 57 kPa was 40 percent of the tensile strength of the aluminum alloy. Probably the first 30 high load levels induced sufficient self (residual) stresses in the test section so as to falsely enhance the fatigue life of the test component and provide overconfidence. All Comet aircraft of this type were taken out of service and additional attention was focused on airframe fatigue design.

Major contributions to the subject of fatigue in the 1950s included the introduction of closed-loop electrohydraulic test systems, which allowed better simulation of load histories on specimens, components, and total mechanical systems. Electron microscopy opened new horizons to better understanding of basic fatigue mechanisms. Irwin* [12] introduced the stress

---

* George R. Irwin (1907). He received his Ph.D. in Physics from the University of Illinois in 1937 and went with the Naval Research Laboratory where he was a research scientist and also became supervisor of the Mechanics Division. Upon retiring from NRL in 1967 he was a professor at Lehigh University until 1972 and then at the University of Maryland. He has continually provided both knowledge and leadership in ASTM committee E-24 on fracture testing. Many ASTM awards plus awards from other professional societies have been bestowed upon him for his contributions to fracture mechanics.

intensity factor K_I, which has been accepted as the basis of linear elastic fracture mechanics (LEFM) and of fatigue crack growth life predictions.

In the early 1960s low cycle strain-controlled fatigue behavior became prominent with the Manson–Coffin [13, 14] relationship between plastic strain amplitude and fatigue life. These ideas are the basis for current notch strain fatigue analysis. Paris [15] in the early 1960s showed that fatigue crack growth rate da/dN could best be described using the stress intensity factor range ΔK_I. In the late 1960s the catastrophic crashes of F-111 aircraft were attributed to brittle fracture of members containing preexisting flaws. These failures, along with fatigue problems in other U.S. Air Force planes, laid the groundwork for the requirements to use fracture mechanics concepts in the B-1 bomber development program of the 1970s. This program included fatigue crack growth life considerations based on a preestablished detectable initial crack size. In 1967 the Point Pleasant Bridge at Point Pleasant, West Virginia collapsed without warning. An extensive investigation of the collapse showed that a cleavage fracture in an eyebar caused by the growth of a flaw to critical size was responsible for the collapse. The initial flaw was due to fatigue, stress corrosion cracking, and/or corrosion fatigue. This failure has had a profound influence on subsequent design requirements established by AASHTO (American Association of State and Highway and Transportation Officials). In July 1974 the U.S. Air Force issued Mil A-83444, which defines damage tolerance requirements for the design of new military aircraft. The use of fracture mechanics as a tool for fatigue was thus thoroughly established through practice and through regulations.

REFERENCES

1. R. E. Peterson, "Discussion of a Century Ago Concerning the Nature of Fatigue, and Review of Some of the Subsequent Researches Concerning the Mechanism of Fatigue,"ASTM Bull., No. 164, Feb. 1950, p. 50.

2. "Wöhler's Experiments on the Strength of Metals," *Engineering*, August 23, 1967, p. 160.

3. H. J. Gough, *The Fatigue of Metals*, Scott, Greenwood and Son, London, 1924.

4. H. F. Moore and J. B. Kommers, *The Fatigue of Metals*, McGraw-Hill Book Co., New York, 1927.

5. A. A. Griffith, "The Phenomena of Rupture and Flow in Solids," *Trans. R. Soc. (Lond.)*, Vol. A221, 1920, p. 163.

6. B. P. Haigh, "The Relative Safety of Mild and High-Tensile Alloy Steels under Alternating and Pulsating Stresses," *Proc. Inst. Automob. Eng.*, Vol. 24, 1929/ 1930, p. 320.

7. J. O. Almen and P. H. Black, *Residual Stresses and Fatigue in Metals*, McGraw-Hill Book Co., New York, 1963.

8. O. J. Horger and T. L. Maulbetsch, "Increasing the Fatigue Strength of Press Fitted Axle Assemblies by Cold Rolling," *Trans. ASME,* Vol. 58, 1936, p. A91.

9. H. Neuber, *Kerbspannungslehre,* Springer-Verlag Berlin, 1937 (in German); *Theory of Notches*, J. W. Edwards, Ann Arbor, Mi., 1946.

10. M. A. Miner, "Cumulative Damage in Fatigue," *Trans. ASME, J. Appl. Mech.,* Vol. 67, Sept. 1945, p. A159.

11. A. Palmgren, "Die Lebensdauer von Kugellagern," ZDVDI, Vol. 68, No. 14, 1924, p. 339 (in German).

12. G. R. Irwin, "Analysis of Stresses and Strains Near the End of a Crack Traversing a Plate," *Trans. ASME, J. Appl. Mech.* Vol. 24, 1957, p. 361.

13. S. S. Manson, Discussion of Ref. 14, *Trans. ASME J. Basic Eng.,* Vol. 84, No. 4, Dec. 1962, p. 537.

14. J. F. Tavernelli and L. F. Coffin, Jr., "Experimental Support for Generalized Equation Predicting Low Cycle Fatigue," *Trans. ASME, J. Basic Eng.,* Vol. 84, No., 4, Dec. 1962, p. 533.

15. P. C. Paris and F. Erdogan, "A Critical Analysis of Crack Propagation Law," *Trans. ASME, J. Basic Eng.,* Vol. 85, No. 4, 1963, p. 528.

CHAPTER 2

FATIGUE DESIGN METHODS

Fatigue design requires two ingredients: a judgment whether there is any danger of fatigue failure and awareness of the factors that may increase the danger or decrease it. How these ingredients are used, checked, and quantified depends on the design task.

2.1 STRATEGIES IN FATIGUE DESIGN

Depending on the purpose of his or her design the designer will proceed along different lines, and he or she will use this book in different ways. For purposes of illustration we look at four different tasks:

1. Designing a device, perhaps a special bending tool or a test rig, to be used in the plant where it was designed. We call it an "in-house tool."
2. Changing an existing product, making it larger or smaller than previous designs, or using different shapes, perhaps a linkage and coil spring in place of a leaf spring. We call it a "new model."
3. Setting up a major project, quite different from past practice. A space craft or an ocean drilling rig or a new type tree harvester might be examples. We call it a "novel product."
4. Designing a highway bridge or a steam boiler. The expected loads, acceptable methods of analysis, and permissible stresses are specified by the customer or by a code authority. We call it "design to code."

2.1.1 The In-House Tool

If a part of the tool is subjected to repeated loads, as, for instance, a ratchet mechanism or a rotating shaft carrying a pulley, it must be designed to avoid fatigue failure. For the in-house tool the information contained in this book is sufficient to do this, provided the designer knows the expected load-time history to which the part will be subjected in service. He or she will sketch

7

a shape that avoids stress concentrations as much as possible, will estimate the stresses, and will select a material and treatment depending on requirements for weight, space, and cost. He or she will use a suitable margin between the stress that corresponds to 50 percent probability of failure at the desired life and the stresses that he or she permits. A second and third iteration may be required to balance the conflicting factors of weight or space, expected life, and cost. If the expected loadings are not uniformly repeated, the designer will consider cumulative damage.

The differences between design for fatigue resistance and design for a few loadings are greater attention to details of shape and treatments, and the need to decide on a required lifetime of the part. The designer will prevent serious consequences of failure by making the part accessible for inspection and replacement, by fail-safe design, or by using large safety factors.

2.1.2 The New Model

For a new model more certainty is required and more data should be available from service records or previous models. In addition to the steps outlined in Section 2.1.1, there usually are tests to confirm the assumptions and calculations. Broken parts from previous models provide the most useful data. They can serve to adjust the test procedures so that testing produces failures similar in location and appearance to service failures.

Tests that produce other types of failures probably have a wrong type of loading or a wrong amplitude. From experience with previous models one also knows sometimes what type of accelerated uniform cycle test is an index of satisfactory performance. In testing passenger car suspension springs, for instance, it has been found that 200,000 cycles of strokes from full rebound to maximum possible deflection was an acceptable test that was much quicker and less expensive than a spectrum test with different amplitudes. Data on loads encountered by the parts may be available directly from previous models or by analogy with previous models. Instead of doing a complete stress analysis it may be possible to determine the relation of significant stresses to loads from measurements on previous satisfactory models and to reproduce the same relation in the new model.

The information in this book will be useful in guiding decisions on design changes when older models are improved or changed.

2.1.3 The Novel Product

This requires the greatest effort in fatigue design. Predicting the future loads is the most important factor. No amount of stress analysis can overcome an erroneous load prediciton. After the loads or load spectra have been obtained one can make analyses of the fatigue worthiness of all parts. They usually are verified by component fatigue tests, which may lead to design modifications. Whenever possible prototypes or pilot models are operated to confirm functional performance and the predicted loads.

2.1.4 Design to Code

Many industries provide data on permissible stresses. The American Welding Society, for instance, has published curves that show recommended stresses as a function of the desired life for various types of welds. Such codes permit the designer to use data based on the experience of many others. As a rule a design according to code is a conservative safe design. In case of a product liability lawsuit U.S. courts do not accept compliance with a code as sufficient to exonerate the maker or seller of a product that eventually failed. Design to code is the only situation in which it may be worthwhile to carry out fatigue calculations to three significant digits. In all other situations the uncertainties are so great that results carried out to ± 5 percent are more precise than warranted by the accuracy of the data.

2.2 FATIGUE DESIGN CRITERIA

Criteria for fatigue design have evolved from so-called infinite life to damage tolerance. Each of the successively developed criteria still has its place, depending on the application.

2.2.1 Infinite-Life Design

Unlimited safety is the oldest criterion. It requires design stresses to be safely below the pertinent fatigue limit. For parts subject to many millions of almost uniform cycles, like engine valve springs, this is still a good design criterion.

2.2.2 Safe-Life Design

Infinite-life design was appropriate for the railroad axles that Wöhler investigated, but automobile designers learned to use parts that, if tested at the maximum expected stress or load, would last only some hundred thousands of cycles instead of many millions. The maximum load or stress in a suspension spring or in a reverse gear may never occur during the life of a car; designing for a finite life under such loads is quite satisfactory. The practice of designing for a finite life is known as "safe-life" design. It is used in many other industries too, for instance, in pressure vessel design and in jet engine design.

The safe life must of course include a margin for the scatter of fatigue results and for other unknown factors. The calculations may be based on stress–life relations, strain–life relations, or crack growth relations.

Ball bearings and roller bearings are noteworthy examples of safe-life design. The ratings for such bearings are often given in terms of a reference load that 90 percent of all bearings are expected to withstand for a given lifetime, for instance, 3000 hours at 500 RPM or 90 million revolutions. For different loads or lives or for different probabilities of failure the bearing manufacturers list conversion formulas. They do not list any load for infinite life, nor for zero probability of failure at any life.

The margin for safety in safe-life design may be taken in terms of life (e.g., calculated life = 20 × desired life), in terms of load (e.g., assumed load = 2 × expected load), or by specifying that both margins must be satisfied, as in the ASME Boiler Code.

2.2.3 Fail-Safe Design

Fail-safe fatigue design criteria were developed by aircraft engineers. They could not tolerate the added weight required by large safety factors nor the danger to lives implied by small safety factors. Fail-safe design recognizes that fatigue cracks may occur and arranges the structure so that cracks will not lead to failure of the structure before they are detected and repaired. Multiple load paths and crack stoppers built at intervals into the structure are some of the means used to achieve fail-safe design. This philosophy originally applied mainly to air frames (wings, fuselages, control surfaces). It is now also used in other applications. Engines are fail-safe only in multiengine planes. A landing gear is not fail-safe but is designed for a safe life.

2.2.4 Damage Tolerant Design

This philosophy is a refinement of the fail-safe philosophy. It assumes that cracks will exist—caused either by processing or by fatigue—and uses fracture mechanics analyses and tests to check whether such cracks will grow large enough to produce failures before they are sure to be detected by periodic inspection. This philosophy looks for materials with slow crack growth and high fracture toughness. Damage tolerant design has been specified by the U.S. Air Force in some contracts. In pressure vessel design "leak before burst" is an expression of this philosophy.

2.3 ANALYSIS AND TESTING

At present, a design based on analysis alone—without testing—either requires large margins for uncertainty or must permit a probability of failure within the range of available data, which usually means 5 percent or more. Parts designed for high rigidity or for static loads much greater than cyclic loads generally have a calculated fatigue life that is an order of magnitude greater than the required life. This may be sufficient without verification by fatigue tests. A probability of failure of a few percent can be permitted if failures do not endanger lives and if replacement is considered a routine matter, as in automobile fan belts. In most other situations analysis needs to be confirmed by tests.

Fatigue testing for design verification, or development testing, is an art, far more demanding than the art of fatigue testing for research, because it requires the engineer to make the test represent conditions of use, but far less demanding than fatigue testing for research in its requirements for precision. Loading and environment similar to those encountered in service are prime requirements for development testing. Determining the service loads may be a major task. The acceleration of development testing poses problems. It is required to bring products to market before the competition or to find a fix for improving marginal products. Increasing loads beyond service loads accelerates tests but may produce misleading results: self stresses (residual stresses) that might have remained in service may be changed by excessive test loads. Fretting can be missed and corrosion may not have time enough to produce its full effect. Eliminating many small load cycles from the test spectrum has been common, but small load cycles must be included to obtain realistic results when most of the fatigue life is spent in propagating cracks.

Test parts should be processed just like production parts because differences in processing (for instance, cut threads instead of rolled threads) may have a major effect on fatigue resistance.

2.4 PROBABILISTIC DESIGN AND RELIABILITY

Fatigue data available at present permit probabilistic design for a few situations down to a probability of failure of about 10 percent. For lower probabilities we hardly ever have the necessary data. Extrapolation of known probability data requires large margins for uncertainty or safety factors. Fatigue reliability can be determined from service experience. Without service experience analyses of designs can predict gross lack of reliability if the design has clear defects, and they are useful for that purpose. Analyses of design without data from service experience cannot, with our present knowledge, provide quantitative reliability figures in the ranges that are required or desired for service.

2.5 INSPECTION AND ACQUISITION OF RELEVANT EXPERIENCE

Imperfections of design will eventually become known. Either the part is too weak and fails too often or it is too strong and a competitor can produce it more economically. Development engineering is an effort to find and fix weaknesses before customers and competition find them. Obtaining records of loads from customers and from proving grounds is one phase of development engineering. Deciding which loads are frequent, which are occasional, which are exceptional, and how much greater loads than those measured can occur is another aspect of development engineering.

Inspection of units in service is also a way to avoid surprises. Many organizations try to put an early production model into severe service with a friendly user and to inspect it very carefully at frequent intervals to find any weaknesses before others find them. Determining suitable inspection intervals and procedures of inspection in service is often part of the fatigue design task. We have seen railway inspectors hit each axle of express trains with long handled hammers to detect fatigue cracks by sound before the cracks become large enough to produce fractures. Airplanes are periodically inspected for cracks. Excessive inspection is wasteful and inspection delayed too long may be fatal. Inspection as part of the production process is not a direct responsibility of the designers, but they mandate it by their specifications. A shoulder without dimension for fillet radius may lead to failure. A specified fillet radius, for instance, 1 ± 0.2 mm, requires inspection, as do full penetration of welds and shot-peening to a specified intensity.

2.6 SUMMARY

The fatigue design process is determined by the features of the design objective and by the extent of the available knowledge. For a few types of products, like engine valve springs or ball bearings, our knowledge is adequate to produce a good economical design if the service loads can be predicted and the service environment is similar to the environment for which experience has been obtained.

For many other situations our theories and algorithms are capable of determining designs only with a large margin for ignorance. A choice between large safety factors and special tests must then be made. The total cost of design, testing, and manufacture must be balanced against the cost (in money, goodwill, reputation or even lives) of fatigue failures. Different design strategies evolve as the weights of these factors vary in different situations.

PROBLEMS FOR CHAPTER 2

1. What safety critical parts on your automobile are: (*a*) fail-safe and (*b*) safe-life?
   How could the critical safe-life parts be made fail-safe? Is this needed?
2. What fatigue design considerations must be made when converting a regular commercial jet aircraft to the "stretch" version?

3. What loading spectra can be assumed for the following situations and how would you determine these loading spectra?

    (*a*) Off-shore oil rig structure.
    (*b*) Short-span highway bridge.
    (*c*) Alaska pipeline.
    (*d*) Motorcycle front axle.
    (*e*) Propeller shaft on oil tanker.

CHAPTER 3

MACRO/MICRO ASPECTS OF FATIGUE OF METALS

By common usage, the word "fatigue" refers to the behavior of materials under the action of repeated stresses or strains as distinguished from the behavior under monotonic or static stresses or strains. The definition of fatigue as currently stated by ASTM follows [1].

> The process of *progressive localized* permanent structural change occurring in a material subjected to conditions which produce fluctuating stresses and strains at some point or points and which may culminate in *cracks* or complete *fracture* after a sufficient number of fluctuations.

Four key words have been italicized in the definition to attract special attention.

The word "progressive" implies the fatigue process occurs over a period of time or usage. A fatigue failure is often very sudden with no external warning; however, the mechanisms involved may have been operating since the beginning of the component or structure usage.

The word "localized" implies that the fatigue process operates at local areas rather than throughout the entire component or structure. These local areas can have high stresses and strains due to external load transfer, abrupt changes in geometry, temperature differentials, self-stresses (residual stresses), and material imperfections. The engineer must be very concerned with these local areas.

The word "cracks" is often the most misunderstood and misused part of fatigue. It somehow seems repulsive or promotes an "I don't believe it" attitude that too many engineers and managers have concerning the role of a crack or cracks in the fatigue process. The ultimate cause of all fatigue failures is that a crack has grown to a point at which the remaining material can no longer tolerate the stresses or strains, and sudden fracture occurs. The crack had to grow to this size because of the cyclic loading. In fact, because of costly fatigue failures in the aerospace fields current aircraft design criteria for certain safety critical parts require the assumption that small cracklike defects exist prior to initial use of the aircraft [2]. The fatigue life of these safety critical parts is thus based solely on crack propagation.

The word "fracture" implies the last stage of the fatigue process is separation of a component or structure into two or more parts.

3.1 FATIGUE FRACTURE SURFACES

Before looking at the microscopic aspects of the fatigue process, we examine a few representative fatigue fracture surfaces. Many of these fracture surfaces have common characteristics, and the words "typical fatigue failure" are often found in the literature and in practice. However, there are many "untypical" fatigue failures too. Both are included in this section.

Figure 3.1 shows a typical fatigue fracture surface of a 97.5 mm (2.48 in.) square thread column from a friction screw press [3]. The thread was not rounded off in the roots, and the flanks exhibited numerous chatter marks, particularly at the lower region A. This poor machining increased the stress concentration in the thread region, which contributed to the fatigue failure. The fracture surface appears to have two distinct regions. The smaller,

FIGURE 3.1 Fatigue failure of a square thread column [3] (reprinted with permission of *Der Maschinenschaden*).

light, somewhat coarse area at the top of the fracture surface is the remaining cross-sectional area that existed at the time of fracture. The other cross-sectional area consists of the fatigue crack (or cracks) region. Many initial cracks can be seen near the lower left outer perimeter, and these are shown by the somewhat radial lines extending around the lower left perimeter. It is this region in which the initial cracking process began. At first the small fatigue cracks propagated at an angle of about 45 degrees for a few millimeters before turning at right angles to the column longitudinal axis, which is the plane of the maximum tensile stress. As the component was subjected to additional cyclic stresses, these small initial cracks grew and joined together such that primarily one fatigue crack propagated across about 80 percent of the surface. Somewhat wavy darker and lighter bands are evident in the main fatigue crack region. These markings are often called beach marks or conchoidal marks. The term "beach marks" has arisen because of the similarity of the fracture pattern to sand markings left after a wave of water leaves a sandy beach. These markings are due to the two adjacent crack surfaces that open, close, and rub together during cyclic loading, and to the crack starting and stopping and growing at different rates during the variable (or random) loading spectrum while interacting with corrosive environment.

If the adjacent fractured parts were placed together (something that should not be done to actual failures since this can obscure metallurgical markings, and thus hamper proper fractographic analysis), they would fit very neatly together and indicate very little permanent deformation. Because of this small permanent visual deformation, fatigue failures are often called "brittle failures." This term, however, should be modified since substantial plastic deformation occurs in small local regions near the fatigue crack tip and at the crack initiation sites. Many fatigue failures do not have appreciable visual permanent deformation, but the mechanisms of crack initiation and propagation involve small local regions of plastic deformations. Failures similar to that in Fig. 3.1 are often called "typical" fatigue failures because they exhibit the following common aspects:

1. Distinct crack initiation site or sites.
2. Beach marks indicative of crack growth.
3. Distinct final fracture region.

Figure 3.2 shows a fatigue failure of an end bearing from an air compressor in a high-capacity locomotive that originated at a metallurgical defect containing slag inclusions and an oxide film near the lower edge [3]. Under the influence of this defect, the high cyclic service stresses led to a fatigue failure. Metallurgical defects, such as inclusions, voids, and so on, are inherent in metals and often act as sites for crack initiation. The top final fracture surface of Fig. 3.2 shows a thin lip around the perimeter. This lip

FIGURE 3.2 Fatigue failure of an end bearing due to slag inclusions [3] (reprinted with permission of *Der Maschinenschaden*).

is often called a shear lip because appreciable permanent deformation can exist here. This shear lip can be quite extensive, very small, or essentially nonexistent. Its occurrence depends on the type of loading and environment and on the ductility of the material. It is more prominent in ductile metals. The total fracture surface shows the crack growth region indicated by the elliptical shaped beach marks, along with the coarse final fracture and shear lip regions.

Figure 3.3a shows a torsion fatigue failure of a 25 mm (1 in.) circular shaft [3]. The fatigue crack initiated at the top surface and propagated along a 45° helix. The crack path is shown schematically in Fig. 3.3c on a 45° plane. This is the plane of maximum tensile stress, which again indicates that fatigue cracks propagate primarily in the plane of the maximum tensile stress. The close-up in Fig. 3.3b shows the smooth fatigue crack shape was semielliptical, a very common surface fatigue crack shape. Note that beach marks are not evident in this smooth semielliptical fatigue crack surface. The final fracture region has a fibrous appearance with radial lines essentially perpendicular to the perimeter of the elliptical fatigue crack. These radial or "river" patterns are often seen on the final fracture surfaces and point to the source of the initial fracture location.

A comparison of fatigue fracture surfaces of cast SAE 0030 steel and hot-rolled SAE 1020 steel, resulting from the same programmed variable amplitude load spectrum using 8.2 mm (0.325 in.) thick keyhole compact type specimens, is shown in Fig. 3.4 [4]. Again, several fatigue cracks have initiated at the keyhole edge, and upon subsequent cycling these initial cracks coalesced into one major crack that occupied most of the crack growth region. Beach marks are not evident on the much coarser cast steel fatigue fracture surface, while a few can be seen at longer crack lengths in the hot-rolled steel. Thus fatigue fracture surfaces can be void of beach marks in both cast and wrought materials.

The major crack growth path for both steels in Fig. 3.4 was flat and

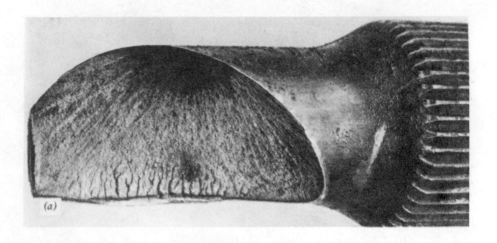

(a)

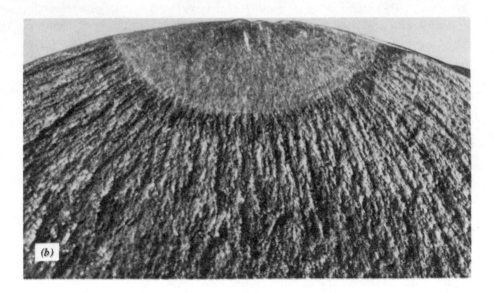

(b)

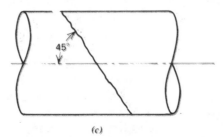

(c)

FIGURE 3.3 Fatigue failure of a torsion shaft [3] (reprinted with permission of *Der Maschinenschaden*). (*a*) Total fracture surface. (*b*) Close-up. (*c*) Idealized form.

double shear as shown in Fig. 3.5b. The mode of fatigue crack growth in sheets or plates is governed primarily by the size of the crack tip plastic zone relative to the sheet thickness. This is principally determined by the yield strength of the material, the magnitude of the applied loads, and the length of the crack.

A general schematic summary of the many types of macroscopic fatigue fracture surfaces is shown in Figs. 3.6 [6] and 3.7. Figure 3.6 shows fracture surfaces due to axial loadings and three types of bending—unidirectional, reversed, and rotating bending. In each case, the fatigue cracks initiate at the surface or corner and propagate in the plane of the maximum tensile stress. For reversed bending, cracks usually initiate at opposite sides, since both sides experience repeated tensile stresses. Figure 3.7 shows the many crack patterns possible in torsion fatigue.

We can summarize the above aspects of fatigue fracture surfaces as follows:

1. The fatigue process involves the initiation and propagation of a crack or cracks to final fracture.
2. The fatigue crack size at fracture can be very small or very large, occupying from less than 1 percent of the fracture surface up to almost 100 percent.
3. Often the fatigue crack region can be distinguished from the final fracture region by beach marks, smoothness, and corrosion. There are many exceptions, however.

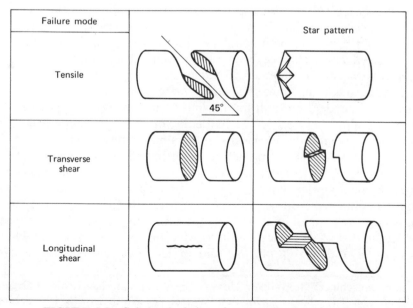

FIGURE 3.7   Schematic fatigue fractures for torsional loads.

4. Fatigue cracks usually initiate at the surface where stresses are highest and where corrosive environment and changes in geometry exist.
5. Macroscopic fatigue cracks usually propagate in the plane of the maximum tensile stress.

3.2 CYCLE DEPENDENT MATERIAL BEHAVIOR

The stress–strain behavior obtained from a monotonic tension or compression test can be quite different from that obtained under cyclic loading. This was first observed during the late nineteenth century by Bauschinger* [7]. His experiments indicated that the yield strength in tension or compression was reduced after applying a load of the opposite sign that caused inelastic deformation. Thus one single reversal of inelastic strain can change the stress–strain behavior of metals. Polakowski and Palchoudhuri [8] determined the effect of many cycles of repeated stresses on the monotonic compression stress–strain relationship of annealed and cold-drawn nickel as shown in Fig. 3.8. They used compression tests rather than tension tests following cyclic loading to eliminate the influence of fatigue cracks if any formed. Cracks would remain closed in a compression test, while in tension they would open, causing both a reduction in the cross-sectional area and sharp discontinuities. Curves *1* and *4* in Fig. 3.8 are the reference annealed and cold-drawn tests, respectively. The magnitude of the maximum cyclic stresses for the annealed condition were higher than the reference yield strength, and hence inelastic deformation occurred in the first cycle. However, the cyclic stresses for the cold-drawn condition were less than the reference yield strength. It is seen that cyclic stressing has shifted the stress–strain curves for both initial conditions. The compression curves for the annealed material have been raised while those for the cold-drawn material have been lowered. Accompanying these changes in the stress–strain properties were similar respective increases or decreases in hardness that were dependent on both the magnitude of the cyclic stresses and the number of applied cycles. Cyclic hardening or cyclic softening is the usual terminology given to these changes in the stress–strain behavior.

Stephens [9] determined the influence of initially elastic cyclic stressing on the tensile yield behavior of iron as shown in Fig. 3.9 (similar behavior occurs for mild steels). The top curve containing the upper and lower yield point was obtained from a monotonic tensile test following 10^6 cycles of alternating stress equal to ±165 MPa (24 ksi). This stress–strain curve was identical to that of the noncycled iron, with the upper and lower yield point stress equal to 290 MPa (42 ksi) and to 234 MPa (34 ksi), respectively.

---

* Johann Bauschinger (1833–1893). He was director of Materials Testing Laboratory and Professor of Mechanics at Munich Polytechnic Institute. In 1884 he organized the first International Congress on Materials Testing.

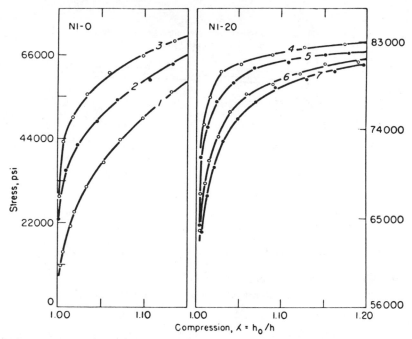

FIGURE 3.8   The effect of alternating tension–compression cyclic stresses on monotonic compression properties of annealed (curves *1–3*) and cold-drawn (curves *4–7*) nickel [8] (reprinted by permission of the American Society for Testing and Materials). (1) As annealed, (2) annealed + 280,000 cycles of ±21,900 psi, (3) annealed + 38,000 cycles of ±30,600 psi, (4) as drawn, (5) drawn + 3,125,000 cycles of ±32,000 psi, (6) drawn + 100,000 cycles of ±43,700 psi, (7) drawn + 237,000 cycles of ±43,700 psi.

However, the upper yield point was completely removed by 75,000 precycles at ±186 MPa (27 ksi) or 35,000 precycles at ±207 MPa (30 ksi). Figures 3.8 and 3.9 thus indicate that cyclic stresses can have a pronounced influence on the overall stress–strain behavior.

   Neither Fig. 3.8 nor 3.9 indicates the progressive changes that take place in the cyclic stress–strain relationship of metals. Continuous monitoring of the stress–strain curve by Morrow [10] during cyclic strain-controlled testing of copper in three initial conditions is shown in Fig. 3.10. These tests were performed on axially loaded specimens in the (*a*) fully annealed condition, (*b*) partially annealed condition, and (*c*) cold-worked condition. The number of applied reversals at different positions of the hysteresis loops are indicated. The components of a hysteresis loop are labeled in Fig. 3.10*a*. The total strain range is denoted by $\Delta\epsilon$ and $\Delta\sigma$ is the stress range. The elastic strain range $\Delta\epsilon_e$ is $\Delta\sigma/E$. By definition,

$$\Delta\epsilon = \Delta\epsilon_p + \Delta\epsilon_e = \Delta\epsilon_p + \Delta\sigma/E \qquad (3.1)$$

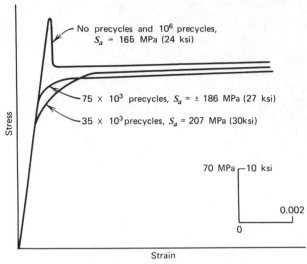

FIGURE 3.9 The effect of cyclic stress on the yield behavior of iron [9] (reprinted by permission of the American Society for Testing and Materials).

where $\Delta\epsilon_p$ is the plastic strain range. The area within a hysteresis loop is energy per unit volume dissipated during a cycle, which is usually in the form of heating. The solid curve from the origin to the first reversal represents the monotonic tensile stress–strain behavior. The last hysteresis loop is also shown solid for ease of comparison. The remaining curves show the appreciable progressive changes in the stress–strain behavior during inelastic cyclic straining. The fully annealed soft copper (a) cyclically hardened, as indicated by an increase in stress range to reach the constant amplitude strain range while the cold-worked copper (c) cyclically softened, as indicated by the decrease in stress to reach a prescribed strain. The partially annealed copper (b) showed initial cyclic hardening followed by cyclic softening. These changes, however, were much less than those of the fully annealed or cold-worked copper. The change in stress amplitude for the constant strain amplitude tests was about 400 percent for the fully annealed copper, 33 percent for the cold-worked copper, and 15 percent for the partially annealed copper. These changes were stabilized or reasonably complete within 10 to 30 percent of the total fatigue life.

A family of stabilized hysteresis loops at different strain ranges is used to obtain the cyclic stress–strain curve for a given material. The tips from the family of multiple loops can be connected, as shown in Fig. 3.11 for Man-Ten steel, to form the cyclic stress–strain curve [11]. This curve does not contain the monotonic upper and lower yield point.

Landgraf et al. [12] obtained both cyclic and monotonic stress–strain curves for several materials shown superimposed in Fig. 3.12. Cyclic softening exists if the cyclic stress–strain curve is below the monotonic curve,

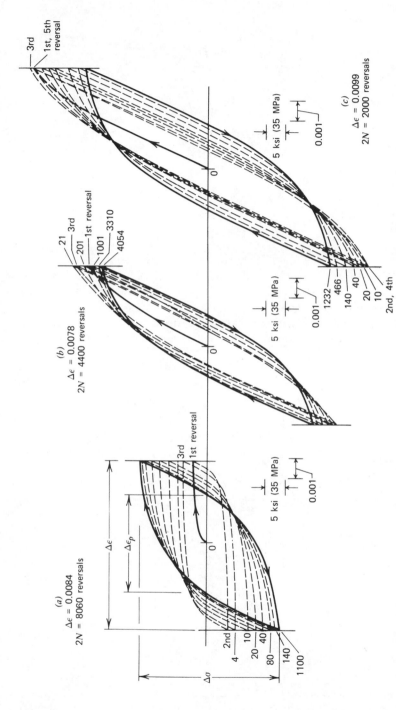

FIGURE 3.10 Stress–strain behavior of copper subjected to cyclic strain-controlled axial loads. (a) Fully annealed showing cyclic hardening. (b) Partially annealed showing small cyclic hardening and softening. (c) Cold-worked showing cyclic softening [10]. (Reprinted by permission of the American Society for Testing and Materials.)

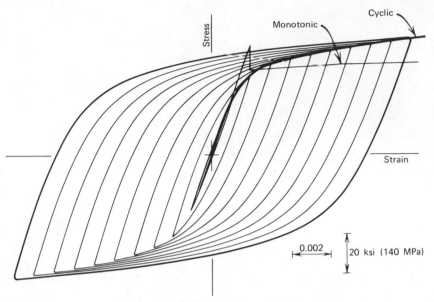

FIGURE 3.11 Step method stable hysteresis loops for determining the cyclic stress–strain curve and comparison with monotonic stress–strain curve, Man-Ten steel [11] (reprinted with permission of the Society of Automotive Engineers).

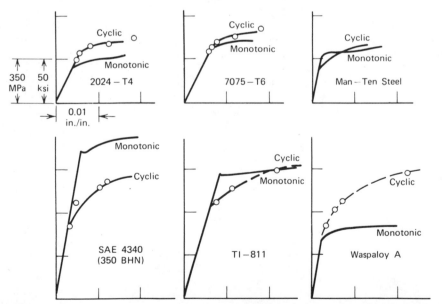

FIGURE 3.12 Monotonic and cyclic stress–strain curves for several materials [12] (reprinted by permission of the American Society for Testing and Materials).

and cyclic hardening is present if it lies above. The difference between the two curves is small in some cases and substantial in others. Manson et al. [13] found that when the ratio of the ultimate tensile strength to the 0.2 percent offset yield strength, S_u/S_y, is greater than 1.4 the material cyclically hardens, and when this ratio is less than 1.2 the material cyclically softens. For ratios between 1.2 and 1.4 it is difficult to predict cyclic changes. From Figs. 3.8, 3.10, and 3.12 and Manson's criteria, it is seen that low strength, soft materials tend to cyclically harden and high strength, hard materials tend to cyclically soften.

The monotonic strain hardening exponent n, given from

$$\sigma = K(\epsilon_p)^n \tag{3.2}$$

where σ and ϵ_p are the true stress and true plastic strain, respectively, usually has a value between 0 and 0.6. If the cyclic stress–strain curve is considered, the exponent n', given from

$$\sigma_a = K' \left(\frac{\Delta\epsilon_p}{2}\right)^{n'} \tag{3.3}$$

where σ_a and $\dfrac{\Delta\epsilon_p}{2}$ are the true stress amplitude and true plastic strain amplitude, respectively, is found to be between about 0.1 and 0.2 for most metals. This range is small compared to the monotonic range.

The macroscopic material behavior given in this section was obtained from uniaxial test specimens with cross-sectional dimension of about 3 to 10 mm (about 1/8 to 3/8 in.). In most cases gross plastic deformation was involved. We should ask the question, how does all this gross plasticity apply to most service fatigue failures where gross plastic deformation does not exist? The answer is quite simple and extremely important. Most fatigue failures begin at local discontinuities where local plasticity exists and crack propagation is governed by local plasticity at the crack tip. The type of behavior shown for gross plastic deformation in Figs. 3.8 to 3.12 is similar to that which occurs locally at notches and crack tips, and thus is extremely important in fatigue of components and structures.

3.3 FATIGUE MECHANISMS

Metals are crystalline in nature, which means atoms are arranged in an ordered manner. Most structural metals are polycrystalline and thus consist of a large number of individual ordered crystals or grains. Each grain has its own particular mechanical properties, ordering direction, and directional properties. Some grains are oriented such that planes of easy slip or glide are in the direction of the maximum applied shear stress. Slip occurs in ductile metals within individual grains by dislocations moving along crys-

tallographic planes. This creates an appearance of one or more planes within
a grain sliding relative to each other. Slip occurs under both monotonic and
cyclic loading. Figure 3.13a shows an edge view of coarse slip normally
associated with monotonic and high stress amplitude cyclic loading. Under
lower stress amplitude cyclic loading fine slip occurs as shown in Fig.
3.13b. Coarse slip can be considered an avalanche of fine movements. Slip
lines shown in Fig. 3.13 appear as parallel lines or bands within a grain
when viewed perpendicular to the free surface. Both fine and coarse slip
are studied with prepolished, etched specimens using optical and electron
microscopy techniques.

Forsyth [14], using electron microscopy, showed that both slip band
intrusions and extrusions occurred on the surface of metals when they were
subjected to cyclic loading. A typical extrusion is shown schematically in
Fig. 3.14 [14]. Slip band intrusions form excellent stress concentrations,
which can be the location for cracks to develop. This slip is primarily
controlled by shear stresses rather than normal stresses. The higher the

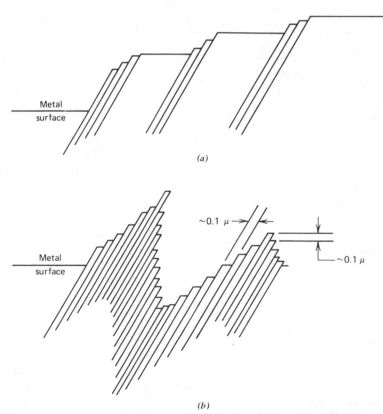

FIGURE 3.13   Slip in ductile metals due to external loads. (a) Static (steady stress).
(b) Cyclic stress.

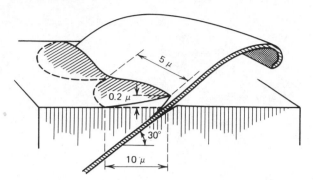

FIGURE 3.14 Extrusion schematic in a slip band [14] (reprinted with permission of Blackie and Son Ltd. Publishing Co.).

shear stress amplitude or the larger the number of repetitions, the greater the slip.

The progressive nature of slip line intensifications as a function of applied cycles for pure polycrystalline nickel is shown in the three stage microphotograph in Fig. 3.15 [15]. Each photograph is from the same microscopic region, but they were taken at 10^4, 5×10^4, and 27×10^4 cycles, respectively. Grain boundaries are in evidence by the closed contours. In Fig. 3.15a, after 10^4 cycles, very little evidence of slip is seen; however, it has occurred in some grains. Figure 3.15b shows a large increase in the number of slip lines after 5×10^4 cycles and also an increased thickness of some slip lines. As cycling continues, more slip lines occur and continue to thicken as shown in Fig. 3.15c. These intensified slip lines or bands are the sources for actual fatigue cracks. Note the slip lines are primarily contained within each grain. A slightly different orientation exists in adjacent grains because of the different directional ordering of each grain. This slip occurs at local regions of high stress and strain, while much of the component or structure may contain very little slip, similar to that in Fig. 3.15a, even at fracture.

The two-dimensional slip lines in Fig. 3.15 are really three dimensional, since they go into the surface to varying depths. Most of the slip bands can be eliminated by removing several microns (0.002 mm) from the surface by electropolishing. However, a few slip bands may become more distinct, and thus they have been called "persistent slip bands." It has been found that fatigue cracks grow from these persistent slip bands [14]. Fatigue life has been substantially improved by removing the persistent slip bands. In fact, intermittent cycling and electropolishing has been employed to cause indefinite life extension such that fatigue failures do not occur [15]. This establishes the fact that early stages of fatigue are primarily a surface phenomenon, and hence the surface plays an important role in fatigue life. Surface effects are emphasized in several later sections.

Fatigue cracks initiate in local slip bands and initially tend to grow in a plane of maximum shear stress range. This growth is quite small, usually of

(a) after 10^4 cycles: general fine slip has developed.

(b) after 5×10^4 cycles : some of the original slip lines have intensified and broadened.

(c) after failure at 27×10^4 cycles: continued development of the intense slip lines has occurred, with some grains and some regions still showing very little evidence of slip.

FIGURE 3.15 The progressive nature of slip in metals subjected to cyclic loading [15] (reprinted with permission of John Wiley and Sons).

the order of several grains. As cycling continues, the fatigue cracks tend to coalesce and grow along planes of maximum tensile stress range, as is described in Section 3.1. The two stages of fatigue crack growth are called stage I and stage II [14].

Fatigue crack growth is shown schematically as a microscopic edge view in Fig. 3.16, where a fatigue crack initiates at the surface and grows across several grains controlled primarily by shear stresses, and then grows in a zigzag manner essentially perpendicular to, and controlled primarily by, the maximum tensile stress range. Most fatigue cracks grow across grain boundaries (transcrystalline) as shown; however, they may also grow along grain boundaries (intercrystalline) but to a much lesser extent. Slip line progression, similar to that in Fig. 3.15, precedes the fatigue crack tip vicinity as the crack grows across the material.

The fatigue mechanism in high strength or brittle metals may not contain slip band formation. Microcracks are often formed directly at discontinuities, such as inclusions or voids, and then grow along planes of maximum tensile stresses.

An electron microscopic analysis of fracture surfaces, such as those in Figs. 3.1 to 3.4, can reveal a wide range of fatigue crack growth mechanisms. Figure 3.17, obtained by Crooker et al. [16], shows three of the more common modes: (a) striation formation, (b) microvoid coalescence, and (c)

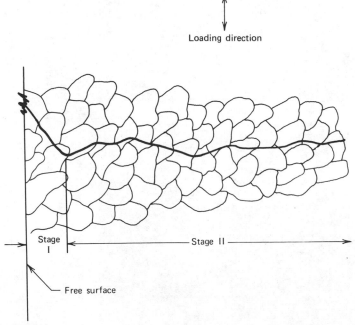

Loading direction

Stage I

Stage II

Free surface

FIGURE 3.16 Schematic of stages I and II transcrystalline microscopic fatigue crack growth.

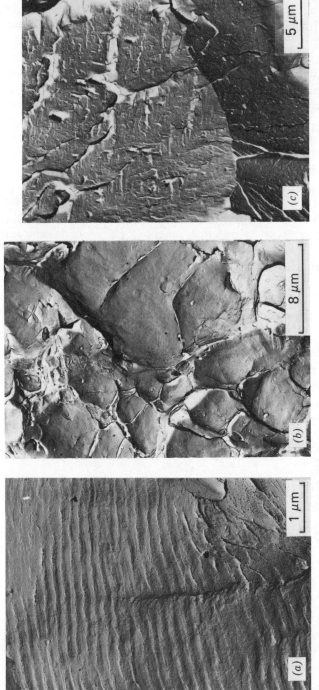

FIGURE 3.17 Examples of microscopic fatigue crack growth in 17-4 PH Stainless steel [16] (reprinted with permission of the American Society for Testing and Materials). (*a*) Striation formation. (*b*) Microvoid coalescence. (*c*) Microcleavage.

microcleavage. Ductile materials often display appreciable striations and microvoid coalescence. The ripples in Fig. 3.17a are called fatigue striations and each striation represents the crack growth in one cycle. These striations are not the beach marks described in Section 3.1. Actually, one beach mark can contain thousands of striations [17]. Electron microscopic magnification between 1000× and 50,000× must be used to view striations. Frequently, they are rather dispersed throughout the fracture surface and may not be seen clearly because of substantial surface rubbing and pounding during repeated loading. They are also difficult to find in high strength materials. Microcleavage crack growth is considered a lower energy process and therefore an undesirable fatigue crack growth mechanism.

The mechanism of fatigue described in this section is summarized schematically in Fig. 3.18 [18]. Slip occurs first, followed by fine cracks that can be seen only at high magnification. These cracks continue to grow under cyclic loading and eventually become visible to the unaided eye. The cracks tend to combine such that just a few major cracks grow. These cracks (or crack) reach a critical size and sudden fracture occurs. The higher the stress magnitude, the sooner all processes occur. Cracks may also stop without

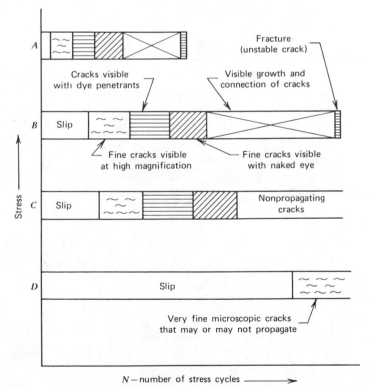

FIGURE 3.18 Schematic representation of the fatigue process [18] (reprinted with permission of McGraw-Hill Book Co.).

further growth as a result of compressive self–stress (residual stress) fields. Thus fatigue consists of crack initiation, propagation, and final fracture. The crack initiation stage depends on the level of thinking. A physicist interested in atomic levels could consider all fatigue as crack growth. A design engineer may think of a crack existing when it can be seen by the unaided eye. The termination of the crack initiation stage is thus quite arbitrary. To avoid this difficulty, the engineer might best think in terms of initiation of microcracks, followed by growth of these microcracks to the macrocrack level, and then macrocrack growth to unstable fracture. Thus a key aspect, no matter which definitions are used, is that fatigue substantially involves the growth of cracks. Only the initiation stage can be absent in a fatigue fracture. Thus we can overcome fatigue by preventing the growth of cracks.

3.4 DOS AND DONT'S IN DESIGN

1. Do recognize that fatigue is a localized progressive behavior involving the initiation and propagation of cracks to final, usually sudden fracture, and that fatigue cracks propagate primarily on planes of maximum tensile stress.
2. Do examine fracture surfaces as part of a postfailure analysis, since substantial information concerning the cause of the fracture can be ascertained. The examination can involve a small magnifying glass or greater magnification up to that of the electron microscope.
3. Don't put fracture surfaces back together again to see if they fit, or allow corrosive environments (including rain and moisture from fingers) to reach the fracture surface. These can obliterate key fractographic details.
4. Do consider that stress–strain behavior at notches or cracks under repeated loading is not the same as that determined under monotonic tensile or compressive loading. Under repeated loading the difference between materials is less compared to monotonic loading.
5. Do take into consideration that your product will very likely contain cracks during its design life time.
6. Do recognize that most fatigue cracks initiate at the surface and therefore surface and manufacturing effects are extremely important.

REFERENCES FOR CHAPTER 3

1. "Standard Definitions of Terms Relating to Fatigue Testing and Statistical Analysis of Data," ASTM Designation E206-72.

2. M. D. Coffin and C. F. Tiffany, "New Air Force Requirements for Structural Safety, Durability, and Life Management," *J. Airc.,* Vol. 13, No. 2, Feb. 1976, p. 93.

3. E. J. Pohl, Ed., *The Face of Metallic Fractures,* Munich Reinsurance Co., Munich, Germany, 1964.

4. R. I. Stephens, P. H. Benner, G. Mauritzson, and G. W. Tindall, "Constant and Variable Amplitude Fatigue Behavior of Eight Steels," *J. Test. Eval.,* Vol. 7, No. 2, March 1979. p. 68.

5. D. Broek, *Elementary Engineering Fracture Mechanics,* Noordhoff International Publishing Co., Leydon, The Netherlands, 1974.

6. G. Jacoby, "Fractographic Methods in Fatigue Research," *Exp. Mech.,* Vol. 5, March 1965, p. 65.

7. J. Bauschinger, "On the Change of the Position of the Elastic Limit of Iron and Steel under Cyclic Variations of Stress," *Mitt. Mech.-Tech. Lab., Munich,* Vol. 13, No. 1, 1886.

8. N. H. Polakowski and A. Palchoudhuri, "Softening of Certain Cold-Worked Metals under the Action of Fatigue Loads," *Proc. ASTM,* Vol. 54, 1954, p. 701.

9. R. I. Stephens, "The Effect of Cyclic Stressing on the Yield Behavior of Vacuum Melted Iron," *J. Mater.,* Vol. 3, No. 2, June 1968, p. 386.

10. J. Morrow, "Cyclic Plastic Strain Energy and Fatigue of Metals," *Internal Friction, Damping, and Cyclic Plasticity,* ASTM STP 378, 1965, p. 45.

11. L. E. Tucker, "A Procedure for Designing Against Fatigue Failure of Notched Parts," SAE paper No. 720265, 1972.

12. R. W. Landgraf, J. Morrow, and T. Endo, "Determination of the Cyclic Stress–Strain Curve," *J. Mater.,* Vol. 4, No. 1, March 1969, p. 176.

13. R. W. Smith, M. H. Hirschberg, and S. S. Manson, "Fatigue Behavior of Materials Under Strain Cycling in Low and Intermediate Life Range," NASA TN D-1574, April 1963.

14. P. J. E. Forsyth, *The Physical Basis of Metal Fatigue,* American Elsevier Publishing Co., New York, 1969.

15. A. J. Kennedy, *Processes of Creep and Fatigue in Metals,* John Wiley and Sons, New York, 1963.

16. T. W. Crooker, D. F. Hasson, and G. R. Yoder, "Micromechanistic Interpretation of Cyclic Crack Growth Behavior in 17-4 PH Stainless Steel," *Fractography—Microscopic Cracking Processes,* ASTM STP 600, 1976, p. 205.

17. R. W. Hertzberg, *Deformation and Fracture Mechanics of Engineering Materials,* John Wiley and Sons, New York, 1976.

18. R. H. Christensen, "Fatigue Cracking, Fatigue Damage, and Their Detection," *Metal Fatigue,* G. Sines and J. L. Waisman, Eds., McGraw-Hill Book Co., New York, 1959, p. 376.

PROBLEMS FOR CHAPTER 3

1. What features distinguish fatigue fracture surfaces from monotonic fractures? When might you expect fatigue fractures to look similar to monotonic fractures with the unaided eye?

2. Two engineers were discussing fatigue failures and one stated, "fatigue failures are brittle" while the other stated, "fatigue failures are ductile." Discuss the pros and cons of the two arguments.

3. Using Table A.2 comment on any possible correlation between the monotonic and cyclic strain hardening exponents n and n'.

4. Why is it important to know if a material cyclically softens or hardens in fatigue?

5. Plot curves of the plastic strain range, $\Delta\epsilon_p$, versus applied reversals, $2N$, for each of the materials in Fig. 3.10. What significance can you ascertain from these curves?

6. Using Equations 3.1 to 3.3 and data from Table A.2 construct the monotonic and cyclic stress–strain curves for RQC-100 hot-rolled steel for strains between 0 and 2 percent. Assume Young's modulus E is 200,000 MPa. Does RQC-100 cyclically harden or soften?

7. How might you differentiate between cyclic slip and an actual micro fatigue crack?

8. What materials and conditions might have fatigue crack growth occur by microcleavage, which is undesirable? How can you avoid this in real structures and materials selection?

9. In line with current product liability litigation, how should we inform our management and legal officers that our products may contain small cracklike flaws or defects before the products leave the factory? Also, provide a list of what these defects might be and what size they might be.

10. List the different nondestructive inspection (NDI) techniques that may aid in measuring crack sizes and also in determining if cracks exist.

CHAPTER 4

FUNDAMENTALS OF LEFM FOR APPLICATION TO FATIGUE CRACK GROWTH AND FRACTURE

The importance of fatigue crack propagation and fracture was brought out in the preceding chapter. This chapter deals with the quantitative methods of handling crack propagation and final fracture. This requires the use of fracture mechanics concepts, with linear elastic fracture mechanics (LEFM) concepts being the most successful. The stress intensity factor K and the stress intensity factor range ΔK are used extensively. Their units are stress times square root of length. We bring out the need for LEFM in the discussion below.

Figure 4.1 shows schematically three crack length versus applied cycles curves for three identical test specimens subjected to different repeated stress levels with $S_1 > S_2 > S_3$. All specimens contained the same initial small crack length, and in each test the minimum stress was zero. We see that with higher stresses the crack propagation rates are higher and the fatigue life is shorter. The crack lengths at fracture were shorter at the higher stress levels. Therefore, total life to fracture depended on the initial crack length, the stress magnitude, and the final fracture resistance of the material. We must ask ourselves, How can fatigue crack growth data, such as that in Fig. 4.1, be used in fatigue design? The format of Fig. 4.1 is not applicable to fatigue design except under the exact same conditions used in obtaining the data. Thus the need arises to apply LEFM concepts to reduce Fig. 4.1 data to a format useful in fatigue design. This involves obtaining crack growth rate, da/dN, versus the applied stress intensity factor range, ΔK, as shown schematically in Fig. 4.2. This sigmoidal shaped curve is essentially independent of initial crack lengths. Stress range, ΔS, and crack length, a, are included in ΔK. Thus, using the proper stress intensity factor for a given component and crack, integration of the sigmoidal shaped curve can provide fatigue crack growth life for components subjected to different stress levels and different initial crack sizes. Crack length at fracture can

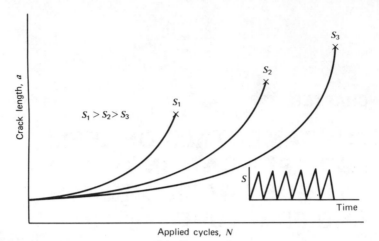

FIGURE 4.1 Fatigue crack length versus applied cycles. Fracture is indicated by ×.

be estimated from critical stress intensity values K_c or K_{Ic}, called fracture toughness. Complex interaction or sequence effects can be considered by analyzing plastic zone sizes at the crack tip along with crack closure models.

This chapter provides an introduction to the important aspects of linear elastic fracture mechanics, which is used in later chapters to describe and predict fatigue crack growth and final fracture. This chapter does not contain the mathematics used to develop the theory but does provide the background fracture mechanics concepts and numerical tools needed for fatigue

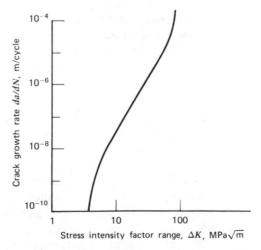

FIGURE 4.2 Schematic sigmoidal behavior of fatigue crack growth rate versus stress intensity factor range.

design involving fatigue crack growth and final fracture. Several excellent textbooks on fracture mechanics can be referred to for greater rigor and detail [1–4]. Fracture mechanics uses the stress intensity factor K, the strain energy release rate G, and the J-integral J, along with critical or limiting values of these terms. We are involved exclusively with the stress intensity factor K and its critical or limiting values K_c. K_{Ic}, and ΔK_{th}. Fracture mechanics has been used heavily in the aerospace, nuclear, and ship industries with only a recent extension to the ground vehicle industry. The chapter is divided into three sections, dealing first with the stress intensity factor K, then plastic zone sizes r_y, followed by fracture toughness K_c and K_{Ic}.

4.1 STRESS INTENSITY FACTOR *K*

Figure 4.3 shows three modes in which a crack can extend. Mode I is the opening mode, which is the most common, particularly in fatigue, and has received the greatest amount of investigation. Mode II is the shearing or sliding mode, and mode III is the tearing or antiplane mode. Combinations of these crack extension modes can also occur, particularly modes I and III as shown in Fig. 3.5. Now let us consider a through thickness sharp crack in a linear elastic isotropic body subjected to Mode I loading. Such a two-dimensional crack is shown schematically in Fig. 4.4. An arbitrary stress element in the vicinity of the crack tip with coordinates r and θ relative to the crack tip and crack plane is also shown. Using the mathematical theory of linear elasticity and the Westergaard stress function in complex form, the stresses at any point near the crack tip can be derived [5]. These stresses are given in Fig. 4.4. Higher order terms exist, but these are negligible in the vicinity of the crack tip. It should be noted that by definition the normal and shear stresses involving the z direction (perpendicular to the x–y plane) are zero for plane stress, while the normal and shear strains (and shear stresses) involving the z direction are zero for plane strain.

Figure 4.4 shows that elastic normal and elastic shear stresses in the vicinity of the crack tip are dependent on r, θ, and K only. The magnitudes

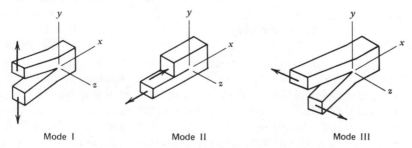

Mode I   Mode II   Mode III

FIGURE 4.3 Three modes of crack extension.

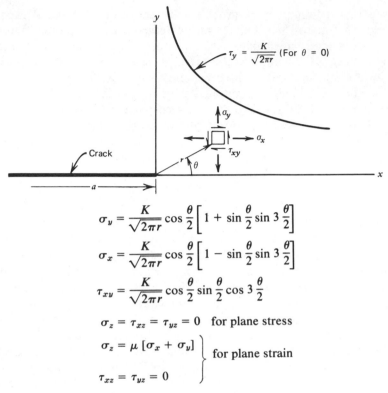

$$\sigma_y = \frac{K}{\sqrt{2\pi r}} \cos \frac{\theta}{2} \left[1 + \sin \frac{\theta}{2} \sin 3 \frac{\theta}{2} \right]$$

$$\sigma_x = \frac{K}{\sqrt{2\pi r}} \cos \frac{\theta}{2} \left[1 - \sin \frac{\theta}{2} \sin 3 \frac{\theta}{2} \right]$$

$$\tau_{xy} = \frac{K}{\sqrt{2\pi r}} \cos \frac{\theta}{2} \sin \frac{\theta}{2} \cos 3 \frac{\theta}{2}$$

$$\sigma_z = \tau_{xz} = \tau_{yz} = 0 \quad \text{for plane stress}$$

$$\left. \begin{array}{l} \sigma_z = \mu \left[\sigma_x + \sigma_y \right] \\[2mm] \tau_{xz} = \tau_{yz} = 0 \end{array} \right\} \quad \text{for plane strain}$$

FIGURE 4.4   Elastic stresses near the crack tip ($r/a \ll 1$).

of these stresses at a given point are thus dependent entirely on K. For this reason, K is called a stress field parameter, or stress intensity factor. K is not to be confused with the elastic stress concentration factor K_t, which is the ratio of the maximum stress at a notch to the nominal stress at the notch. The value of the stress intensity factor, K, depends on the loading, body configuration, crack shape, and mode of crack displacement. K used without a mode subscript I, II, or III normally refers to mode I.

The elastic stress distribution in the y direction for $\theta = 0$ is shown in Fig. 4.4. As r approaches zero, the stress at the crack tip approaches infinity, and thus a stress singularity exists at $r = 0$. Since infinite stresses cannot exist in a physical body, the elastic solution must be modified to account for crack tip plasticity. If, however, the plastic zone size r_y at the crack tip is small relative to local geometry (for example, r_y/t and $r_y/a \gtrsim 0.1$ where t is thickness), little or no modification to the stress intensity factor, K, is needed. Thus an important restriction to the use of linear elastic fracture mechanics is that the plastic zone size at the crack tip must be small relative to the geometrical dimensions of the specimen or part. A definite limiting condition for linear elastic fracture mechanics is that nominal stresses in the crack plane must be less than the yield strength. In actual usage the nominal stress in the crack plane should be less than 0.8 times the yield strength [6].

Values of K for various loadings and configurations can be calculated using the theory of elasticity involving both analytical and numerical calculations along with experimental methods. The most common reference value of K is for a two-dimensional center crack of length $2a$ in an infinite sheet subjected to a uniform tensile stress S. For the infinite sheet, K is:

$$K = S\sqrt{\pi a} \approx 1.77 \, S\sqrt{a} \tag{4.1}$$

Units of K are MPa$\sqrt{m}$ and ksi$\sqrt{in.}$, where 1 Mpa$\sqrt{m}$ = 0.91 ksi$\sqrt{in.}$

The stress intensity factor for other crack geometries, configurations, and loadings are usually modifications of Eq. 4.1, such that

$$K = S\sqrt{\pi a}\,\alpha \quad \text{or} \quad S\sqrt{\pi a}\,f\left(\frac{a}{w}\right) \quad \text{or} \quad S\sqrt{a}\,Y \tag{4.2}$$

where α, $f(a/w)$, and Y are dimensionless parameters, w is a width dimension, and S is the nominal stress, assuming the crack did not exist. For central cracks, the crack length is taken as $2a$ (or $2c$) and for edge cracks the crack length used is just a (or c). Opening mode I stress intensity expressions in the form of dimensionless curves are given in Fig. 4.5 for several common configurations of thickness B [2, 7]. In each case it is evident that K depends on the crack length to width ratio, a/w. These tabulations were obtained from analytical or numerical solutions, often in the form of polynomials. The actual mathematical expression is of greater importance in fatigue crack growth since numerical integration is usually required. Several of these common mathematical expressions are given in Table 4.1. For the single or double edge crack in a semiinfinite plate $(a/w \rightarrow 0)$

$$K = 1.12 \, S\sqrt{\pi a} \approx 2S\sqrt{a} \tag{4.3}$$

where 1.12 is the free edge correction. Additional stress intensity factor expressions for all three modes, K_I, K_{II}, and K_{III} can be found in references 1 to 10. Superposition of K expressions can also be used for each separate mode.

The elliptical crack approximates many cracks found in engineering components and structures and has received widespread analytical, numerical, and experimental analysis. Common circular and elliptical embedded and surface cracks are shown in Fig. 4.6. The general reference specimen is the embedded elliptical crack in an infinite body subjected to uniform tension S at infinity (Fig. 4.6c). For this embedded configuration K is:

$$K = \frac{S\sqrt{\pi a}}{\Phi}\left[\sin^2\beta + \left(\frac{a}{c}\right)^2\cos^2\beta\right]^{1/4} \tag{4.4}$$

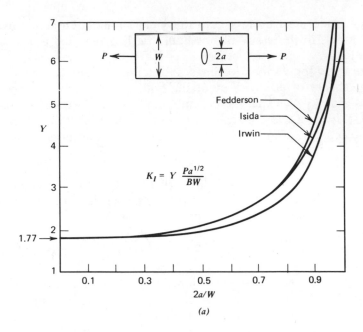

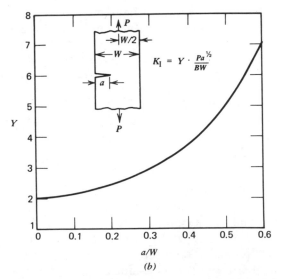

FIGURE 4.5 K_1 for common configurations [2, 7] (reprinted by permission of Noordhoff International Publishing Co. and the American Society for Testing and Materials). (*a*) Center cracked plate in tension [2]. (*b*) Single edge crack in tension [7]. (*c*) Double edge crack in tension [7]. (*d*) Single edge crack in bending [7].

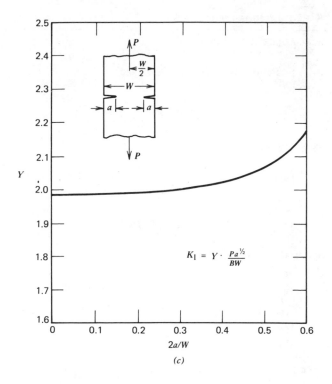

$$K_1 = Y \cdot \frac{P a^{\frac{1}{2}}}{BW}$$

(c)

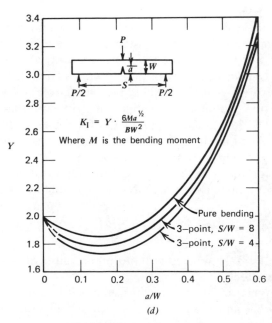

$$K_1 = Y \cdot \frac{6 M a^{\frac{1}{2}}}{BW^2}$$

Where M is the bending moment

Pure bending

3—point, $S/W = 8$

3—point, $S/W = 4$

(d)

FIGURE 4.5 (*Continued*)

43

TABLE 4.1 *K* Expressions for Fig. 4.5

a. Center cracked plate in tension

$$\text{Fedderson: } K_\mathrm{I} = S\sqrt{\pi a}\left[\sec\frac{\pi a}{w}\right]^{1/2}$$

$$\text{Irwin: } K_\mathrm{I} = S\sqrt{\pi a}\left[\frac{w}{\pi a}\tan\frac{\pi a}{w}\right]^{1/2}$$

b. Single-edge crack in tension

$$K_\mathrm{I} = S\sqrt{a}\left[1.99 - 0.41\left(\frac{a}{w}\right) + 18.7\left(\frac{a}{w}\right)^2 - 38.48\left(\frac{a}{w}\right)^3 + 53.85\left(\frac{a}{w}\right)^4\right]$$

where
$$1.12\sqrt{\pi} = 1.99$$
$$S = P/Bw$$

c. Double-edge crack in tension

$$K_\mathrm{I} = S\sqrt{a}\left[1.98 + 0.36\left(\frac{2a}{w}\right) - 2.12\left(\frac{2a}{w}\right)^2 + 3.42\left(\frac{2a}{w}\right)^3\right]$$

where
$$S = P/Bw$$

d. Pure bending of a beam

$$K_\mathrm{I} = \frac{6M\sqrt{a}}{Bw^2}\left[1.99 - 2.47\left(\frac{a}{w}\right) + 12.97\left(\frac{a}{w}\right)^2 - 23.17\left(\frac{a}{w}\right)^3 + 24.8\left(\frac{a}{w}\right)^4\right]$$

where β is the angle shown in Fig. 4.6g, $2a$ is the minor diameter, and $2c$ is the major diameter. The term Φ is the complete elliptical integral of the second kind and depends on the crack aspect ratio a/c. Values of Φ are given in Fig. 4.6h, where it is seen that Φ varies from 1.0 to 1.571 for a/c ranging from zero (very shallow ellipse) to one (circle). K varies along the elliptical crack tip according to the trigonometric expression involving the angle β (Eq. 4.4). The maximum value of K for the embedded crack exists at the minor axis and the minimum is at the major axis. Since fatigue crack growth depends principally on K, the embedded elliptical crack subjected to uniform tension tends to grow to a circle with a uniform K at all points on the crack tip perimeter. For the circular embedded crack

$$K = S\sqrt{\pi a}\left(\frac{2}{\pi}\right) = 2S\sqrt{\frac{a}{\pi}} \approx 1.13\, S\sqrt{a} \qquad (4.5)$$

Surface elliptical cracks tend to grow to other elliptical shapes because of the free surface effect.

The surface semicircular or semielliptical crack in a finite thickness solid (Figs. 4.6d to 4.6f) and the quarter-elliptical corner crack (Figs. 4.6e and

4.6*f*) are very common in fatigue and are extremely more complex compared to the embedded crack in an infinite solid. *K* at the surface intersection and around the crack tip perimeter has been estimated using numerical methods [11–14] and three-dimensional photoelasticity [15]. References 1 to 6 and 8 to 10 also contain *K* estimations for these three dimensional cracks. A general expression for the mode I semielliptical surface crack in a finite

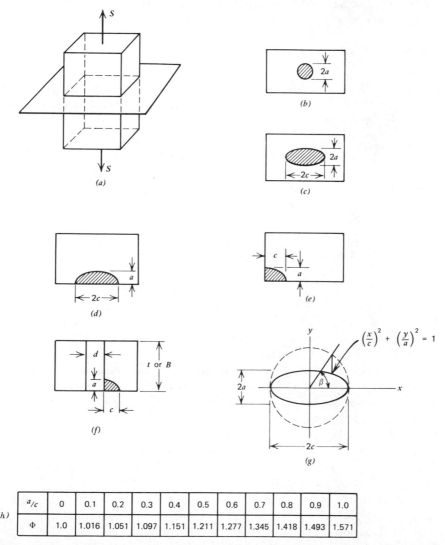

a/c	0	0.1	0.2	0.3	0.4	0.5	0.6	0.7	0.8	0.9	1.0
Φ	1.0	1.016	1.051	1.097	1.151	1.211	1.277	1.345	1.418	1.493	1.571

FIGURE 4.6 Elliptical and circular cracks. (*a*) Tensile loading and crack plane. (*b*) Embedded circular crack. (*c*) Embedded elliptical crack. (*d*) Surface half-elliptical crack. (*e*) Quarter-elliptical corner crack. (*f*) Quarter-elliptical corner crack emananting from a hole. (*g*) Elliptical crack parameters. (*h*) Values of Φ.

thickness plate (Fig. 4.6d) is

$$K = \frac{S\sqrt{\pi a}}{\Phi} M_f M_b \left[\sin^2 \beta + \left(\frac{a}{c}\right)^2 \cos^2 \beta \right]^{1/4} \qquad (4.6)$$

where M_f is a front face correction factor and M_b is a back face correction factor. M_f and M_b are functions of β.

For a surface semielliptical crack in a thick plate K at the deepest point is approximated as:

$$K \simeq \frac{1.12\ S\sqrt{\pi a}}{\Phi} \qquad (4.7)$$

where $M_f \simeq 1.12$ is analogous to the free edge correction of the single edge crack. For the quarter-circular corner crack ($a/c = 1$) in Fig. 4.6e, with two free edges, K is approximated as:

$$K \simeq \frac{(1.12)^2\ S\sqrt{\pi a}}{\Phi} \simeq (1.12)^2\ 2S\sqrt{\frac{a}{\pi}} \simeq 2.5\ S\sqrt{\frac{a}{\pi}} \simeq 1.41\ S\sqrt{a} \qquad (4.8)$$

These simple approximations agree quite well with more complex calculations.

4.2 CRACK TIP PLASTIC ZONE SIZE

Whether a fracture occurs in a ductile or brittle manner, or a fatigue crack grows under cyclic loading, the local plasticity at the crack tip controls both fracture and crack growth. It is possible to calculate a plastic zone size at the crack tip as a function of stress intensity factor, K, and yield strength, S_y, by using the stress equations in Fig. 4.4 and the von Mises or maximum shear stress yield criteria. The resultant monotonic plastic zone shape for mode I using the von Mises criterion is shown schematically in Fig. 4.7. For plane stress conditions, a much larger plastic zone exists compared to plane strain conditions, as indicated in the lower right corner. This is due to σ_z having a different value for plane stress than for plane strain as given in Fig. 4.4, which decreases the magnitude of two of the three principal shear stresses. Now let us assume a through crack exists in a thick plate. The plate's free surfaces have zero normal and shear stresses, and therefore the free surfaces must be in a plane stress condition. However, the interior region of the plate near the crack tip is closer to plane strain conditions as a result of elastic constraint away from the crack. Thus the plastic zone along the crack tip varies similarly to that shown schematically in Fig. 4.7. The actual stress–strain distribution within the plastic zone is difficult to obtain; however, this is not significant for the use of linear elastic facture mechanics in design.

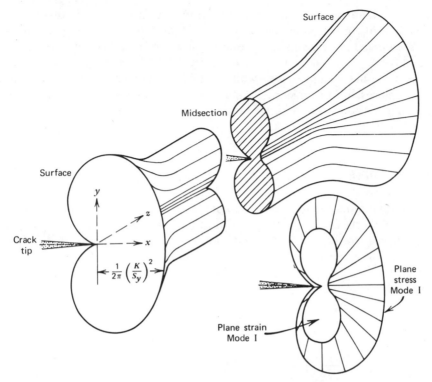

FIGURE 4.7   Plastic zone size at the tip of a through thickness crack.

It is seen in Fig. 4.7 that the plastic zone is proportional to the square of the ratio of the stress intensity factor to the yield strength. Because of plastic relaxation of the stress field in the plastic zone, the actual plane stress plastic zone size is approximately twice the value shown in Fig. 4.7. The plane strain plastic zone size in the plane of the crack is usually taken as one-third the plane stress value. Thus, under monotonic loading, the plane stress plastic zone size, $2r_y$, at the crack tip, in the plane of the crack, is:

$$2r_y = 2\left[\frac{1}{2\pi}\left(\frac{K}{S_y}\right)^2\right] = \frac{1}{\pi}\left(\frac{K}{S_y}\right)^2 \tag{4.9}$$

and for plane strain

$$2r_y \simeq \frac{1}{3\pi}\left(\frac{K}{S_y}\right)^2 \tag{4.10}$$

Additional models for plastic zone size and shape, which have also received wide spread use, have been formulated by Dugdale [16] and Hahn and Rosenfield [17]. Under cyclic loading, a reversed plastic zone occurs when the tensile load is removed. Rice [18] showed that it is much smaller than the peak monotonic values given by Eq. 4.9 or 4.10. For example, the size of the reversed plastic zone after unloading to zero load is only one-quarter of that which existed at the peak tensile load. Then the stresses in the plastic zone are compressive, while outside the plastic zone they change from compression to tension. Fatigue load sequence effects are caused by these plastic zones containing compressive or tensile stresses, as is dis- cussed in later chapters.

4.3 FRACTURE TOUGHNESS—K_c, K_{Ic}

Quantitative monotonic fracture toughness can be obtained for brittle ma- terials from tests using specimens with fatigue cracks with known K expres- sions such as those in Fig. 4.5 and Table 4.1. Critical values of K refer to the condition when a crack extends in a rapid (unstable) manner without an increase in load or applied energy. Critical values of K are denoted with a subscript c as follows:

$$K_c = S_c\sqrt{\pi a_c}f\left(\frac{a_c}{w}\right) \tag{4.11}$$

where S_c is the applied nominal stress at crack instability and a_c is the crack length at instability; K_c is called fracture toughness and depends on the material, temperature, strain rate, environment, thickness, and to a lesser extent, crack length. If K_c is known for a given material and thick- ness, and K is known for a given component and loading, a quantitative design criteria to prevent brittle fracture exists involving applied stress and crack size. Since K_c also represents the stress intensity factor at the last cycle of a fatigue fracture, it can be used to obtain critical crack sizes for fracture under cyclic loading. In 1920 Griffith [19], using elastic energy balance equations and experimentation with brittle glass, was the first to produce a quantitative relationship for brittle fracture of cracked bodies.

The general relationship between fracture toughness, K_c, and thickness is shown in Fig. 4.8. The fracture appearance accompanying the different thicknesses is also shown schematically for single-edge notch specimens. The beach markings at the crack tip represent fatigue precracking at low cyclic stress intensity factor range to assure a sharp crack tip. The fracture toughness values would be higher for dull, or notch type, crack fronts. It is seen that thin parts have a high value of K_c accompanied by appreciable "shear lips" or slant fracture. As the thickness is increased, the percentage of "shear lips" or slant fracture decreases, as does K_c. This type of fracture appearance is called mixed mode. For thick parts, essentially the entire fracture surface is flat and K_c approaches an asymptotic minimum value. Further increase in thickness does not decrease the fracture toughness, nor does it alter the fracture appearance. The minimum value of fracture tough-

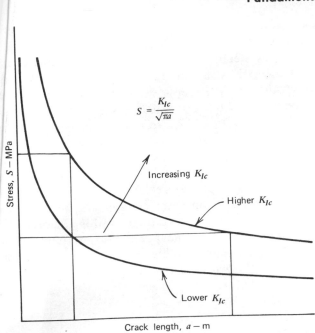

FIGU4.11 Influence of fracture toughness on allowable stress or crack size.

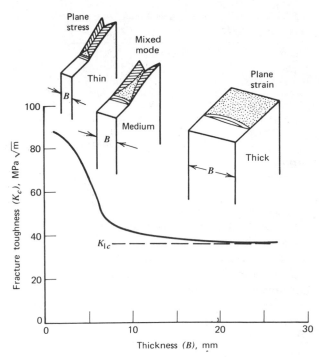

FIGURE 4.8 Effect of specimen thickness on fracture toughness.

otonicading of components containing cracks, a higher fracture toughness results larger allowable crack sizes or larger allowable stresses at fracture.

4.4 DOS AND DON'TS IN DESIGN

1. Do recognize that LEFM can aid both qualitatively and quantitatively in estimating fatigue crack growth life and final fracture. The stress intensity factor K describes the stress field at the tip of a fatigue crack.

2. Do consider that fracture toughness depends much more on metallurgical discontinuities and impurities than does ultimate or yield strength. Low impurity alloys have better fracture toughness.

3. Don't expect that doubling the thickness or doubling the ultimate strength of a component will double the fracture load. Cracks can exist and fracture toughness may drop appreciably with both thickness and ultimate strength increases.

4. Do determine reasonable estimates of stress intensity factors K for components containing cracks by modeling these cracks with more simplified K solutions available in handbooks or through numerical, analytical, or experimental methods.

5. Don't neglect the importance of nondestructive flaw or crack inspection for both initial and periodic inspection periods.

ness is called the "plane strain fracture toughness" K_{Ic}. The subscript I refers to the fact that these fractures occur almost entirely by mode I crack opening. The term "plane strain" is incorporated here since flat fractures best approach a true plane strain constraint throughout most of the crack tip region. For thin sections where appreciable "shear lips" occur, the crack tip region most closely experiences a plane stress situation. Thus plastic zone sizes at fracture are much larger in thin parts as compared to thick parts. Plane strain fracture toughness K_{Ic} is considered a true material property because it is independent of thickness. Approximate thickness required for steels and aluminums to obtain valid K_{Ic} values are given in Table 4.2. Low strength, ductile materials are subject to plane strain fracture at room temperature only if they are very thick. Therefore, most K_{Ic} data have been obtained for the medium and higher strength materials or for the lower strength materials at low temperatures.

A general trend of K_{Ic} at room temperature, as a function of yield strength for three major base alloys, aluminum, titanium, and steel, is given in Fig. 4.9 [20]. As can be seen, a wide range of K_{Ic} can be obtained for a given base alloy. However, a higher yield or ultimate strength generally produces a decrease in K_{Ic} for all materials, and thus a greater susceptibility for catastrophic fracture. This is an important conclusion that too many engineers overlook. Figure 4.9 does not provide K_{Ic} data for all levels of yield

TABLE 4.2   Approximate Thickness Required for Valid K_{Ic} Tests

Steel S_y, MPa (ksi)	Aluminum S_y, MPa (ksi)	Thickness, mm (in.)
690 (100)	275 (40)	> 76 (3)
1030 (150)	345 (50)	76 (3)
1380 (200)	448 (65)	45 (1-¾)
1720 (250)	550 (80)	19 (¾)
2070 (300)	620 (90)	6 (¼)

strength because of the large thickness required for low strength materials. The use of linear elastic fracture mechanics is not suitable for these low strength materials under monotonic loading because of the extensive plastic zones occurring at the crack tip. Exceptions occur, however, at very low temperatures, in the presence of corrosive environments, and under fatigue conditions where only small scale yielding occurs near the crack tip. Appreciable K_{Ic} data for specific materials can be found in reference 21 and some representative K_{Ic} values are given in Table A.3. Much variability, however, occurs for a given yield strength, depending on the type and quality of the material. This is best illustrated for steels in Fig. 4.9, where vacuum-induction melting (VIM) plus vacuum-arc melting (VAR) of the base material produce greater K_{Ic} for a given yield strength than just single vacuum-arc remelting or air melting of steels (AIR). Thus low impurity materials provide better fracture toughness.

4.3   Fracture Toughness—K_c, K_{Ic}

Fracture toughness K_{Ic} of metals is also depende rate, and corrosive environment. Figure 4.10 sh results for a low alloy steel [22]. As the temperatur decreases, while the yield strength increases. Thus or uncracked tensile strength increases with decre flaw or crack resistance can be drastically reduce tends to cause changes in K_{Ic} similar to that of decre That is, higher strain rates often produce lower fra hence, greater crack sensitivity. Corrosive environmer term fracture toughness may show small or large cha rosive environments can cause appreciable decrease under long term stress corrosion cracking and fatigue tions. Environmental considerations are covered in Ch

A general schematic drawing of how changes in fractu ence the relationship between allowable nominal stress a size is shown in Fig. 4.11, which is a plot of

$$S = \frac{K_{Ic}}{\sqrt{\pi a}}$$

This equation comes from equating the stress intensity fact crack in a wide plate to the fracture toughness K_{Ic}. The al the presence of a given crack size is directly proportional toughness, while the allowable crack size for a given stress to the square of the fracture toughness. Thus increasing larger influence on allowable crack size than on allowable st

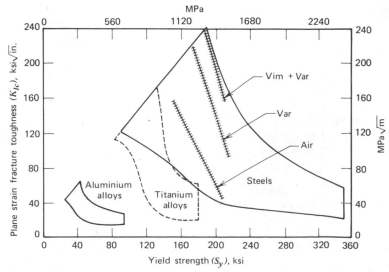

FIGURE 4.9   Locus of plane strain fracture toughness versus yield strength [20].

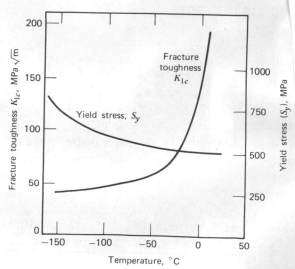

FIGURE 4.10   Variation of K_{Ic} with temperature for low alloy nucl vessel steel A533B [22] (reprinted with permission of E. T. Wessel).

REFERENCES FOR CHAPTER 4

1. J. F. Knott, *Fundamentals of Fracture Mechanics,* Halsted Press, John Wiley and Sons, New York, 1973.

2. D. Broek, *Elementary Engineering Fracture Mechanics,* Noordhoff International Publishing Co., Leyden, The Netherlands, 1974.

3. S. T. Rolfe and J. M. Barsom, *Fracture and Fatigue Control in Structures— Applications of Fracture Mechanics,* Prentice-Hall, Englewood Cliffs, NJ, 1977.

4. R. W. Hertzberg, *Deformation and Fracture Mechanics of Engineering Materials,* John Wiley and Sons, New York, 1976.

5. V. Weiss and S. Yukawa, "Critical Appraisal of Fracture Mechanics," *Fracture Toughness Testing and Its Applications,* ASTM STP 381, 1965, p. 1.

6. *Fracture Toughness Testing and Its Application,* ASTM STP 381, 1965.

7. W. F. Brown, Jr. and J. E. Srawley, *Plane Strain Crack Toughness Testing of High Strength Metallic Materials,* ASTM STP 410, 1966, p. 1.

8. H. Tada, P. C. Paris, and G. R. Irwin, *The Stress Analysis of Cracks Handbook,* Del Research Corp., Hellertown, PA, 1973.

9. G. C. Sih, *Handbook of Stress Intensity Factors,* Institute of Fracture and Solid Mechanics, Lehigh University, Bethleham, PA, 1973.

10. D. P. Rooke and D. J. Cartwright, *Compendium of Stress Intensity Factors,* Her Majesty's Stationery Office, London, 1976.

11. R. C. Shah and A. S. Kobayashi, "Stress Intensity Factors for an Elliptical Crack Approaching the Surface of a Semi-Infinite Solid," *Int. J. Fract.,* Vol. 9, No. 2, 1973, p. 133.

12. F. W. Smith and D. R. Sorenson, "The Semi-Elliptical Surface Crack—A Solution by the Alternating Method," *Int. J. Fract.,* Vol. 12, No. 1, 1976, p. 47.

13. T. A. Cruse and G. J. Meyers, "Three-Dimensional Fracture Mechanics Analysis," *J. Struct. Div. ASCE,* Vol. 103, No. ST2, Feb. 1977, p. 309.

14. D. M. Tracy, "3-D Elastic Singularity Element for Evaluation of K along an Arbitrary Crack Front," *Int. J. Fract.,* Vol. 9, No. 3, 1973, p. 340.

15. C. W. Smith, "Fracture Mechanics," *Experimental Techniques in Fracture Mechanics,* Vol. 2, A. S. Kobayashi, Ed., The Iowa State University Press, 1975, p. 3.

16. D. S. Dugdale, "Yielding of Steel Sheets Containing Slits," *J. Mech. Phys. Solids,* Vol. 8, 1960, p. 100.

17. G. T. Hahn and A. R. Rosenfield, "Local Yielding and Extension of a Crack under Plane Stress," *Acta Metall.,* Vol. 13, No. 3, 1965, p. 293.

18. J. R. Rice, "Mechanics of Crack Tip Deformation and Extension by Fatigue," *Fatigue Crack Propagation,* ASTM STP 415, 1967, p. 247.

19. A. A. Griffith, "The Phenomena of Rupture and Flow in Solids," *Philos. Trans., R. Soc. Lond., Ser. A.,* Vol. 221, 1920, p. 163.

20. W. S. Pellini, "Criteria for Fracture Control Plans," NRL Report 7406, May, 1972.

21. *Damage Tolerant Design Handbook, A Compilation of Fracture and Crack Growth Data for High Strength Alloys,* MCIC-HB-01, Metals and Ceramics Information Center, Battelle, Columbus, OH.

22. E. T. Wessel, "Variation of K_{Ic} with Temperature for Low Alloy Nuclear Pressure Vessel Steel A533B," *Practical Fracture Mechanics for Structural Steel,* Chapman and Hall, 1969.

PROBLEMS FOR CHAPTER 4

1. Using the principle of superposition, determine K_1 for the member shown, which has an offset force P equal to 100 kN. The crack is a through crack. Repeat the calculations for $a = 30$ mm.

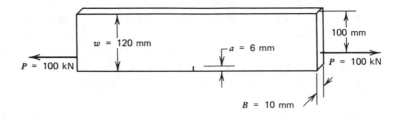

2. Superimpose a plot of K_1 versus angle β (0–90 degrees) for an elliptical embedded crack with $a/c = 1$ (circle), 0.5, and 0.1 (long shallow crack) assuming S and a are constant. Comment on the influence of the crack shape on K_1 for a given stress and crack length a.

3. Compare the approximate K_1 value for a circular corner crack with the results of problem 2.

4. Assuming the A533B steel of Fig. 4.10, what force will cause fracture at 0° and −50°C for a circular corner crack of radius $a = 5$ mm in a thick axially loaded 50 mm × 100 mm cross-section bar? What plastic zone sizes exist at the two fracture conditions? Is the LEFM model justifiable in this problem? What force would cause general yielding if the crack were not present? Discuss the significance of the above questions and answers.

5. An 18 Ni (200) maraging steel thick rod fractured under an axial load. It was suggested to increase the strength and use an 18 Ni (300) maraging steel. Using Table A.3 comment on the above suggestion for (*a*) no cracks present and (*b*) a crack present. For a given crack size, what is the ratio of the load carrying capacity of the two steels? How would you overcome the original fracture problem assuming the crack was present at fracture?

6. Assume the fracture toughness data of Fig. 4.8 and that a uniaxially loaded part is 2.5 mm thick. It is desired to increase the fracture load capacity by a factor of 3. If the thickness were tripled and no cracks

existed, this could be a satisfactory solution. However, if a through crack exists in both parts, would tripling the thickness give the desired tripling in fracture load? What load change at fracture would be achieved?

7. Determine an approximate allowable centrally located through the thickness crack size, $2a$, in a very wide, thick plate of 2024-T3 and 7075-T6 aluminum alloy subjected to a nominal stress of 270 MPa. Repeat the calculation for nominal stresses equal to $S_y/2$. What significance is shown by the above?

CHAPTER 5

CONSTANT AMPLITUDE
FATIGUE TESTS AND DATA

5.1 MONOTONIC UNIAXIAL TENSION/COMPRESSION

Monotonic tension or compression stress–strain properties are usually reported in handbooks and are called for in many specifications. They are easy tests to perform and provide information that has become conventionally accepted. However, their relation to fatigue behavior may be rather remote, as noted in the following excerpt from the preface of a small book on fatigue written by Spangenburg in 1876 [1].

> Spangenburg's experiments, given in the following treatise, were, as will be seen, in continuation of Wöhler's. The results of these very important experiments have been before the profession for some years, but strange to say, seem to have attracted no attention; and tests of iron and steel still go on for the purpose of determining their elasticity, their elongation under strain, their ultimate strength and other qualities, while Wöhler and Spangenburg's experiments show that it is very doubtful that these bear any proportion to the durability of the metals.

This excerpt is still pertinent to modern engineers, who sometimes like to think that fatigue behavior can be predicted from simple tests, such as the monotonic tensile test. This section provides a brief description of the monotonic uniaxial test needed to better prepare for fatigue behavior determination.

Monotonic uniaxial stress–strain behavior can be based on "engineering" stress–strain or on "true" stress–strain relationships. The nominal engineering stress, S, in a uniaxial test specimen is:

$$S = \frac{P}{A_0} \tag{5.1}$$

where P is the axial force and A_0 is the original cross-sectional area. The true stress, σ, in this same test specimen is given by

$$\sigma = \frac{P}{A} \qquad (5.2)$$

where A is the instantaneous cross-sectional area. The engineering strain, e, is given by

$$e = \frac{\Delta l}{l_0} \qquad (5.3)$$

where Δl is the change of length of the original gage length l_0. The true strain, ϵ, is given by

$$\epsilon = \int_{l_0}^{l} \frac{dl}{l} = \ln\left(\frac{l}{l_0}\right) \qquad (5.4)$$

For small strains, less than about 2 percent, the "engineering" stress, S, is approximately equal to the "true" stress, σ, and the "engineering" strain, e, is approximately equal to the "true" strain, ϵ. No distinction between "engineering" and "true" components is needed for these small strains. However, for larger strains the differences become appreciable. With large inelastic deformations a constant volume condition can be assumed such that up to necking $Al = A_0 l_0$. The following relationships among S, σ, e, and ϵ can then be derived.

$$\sigma = S(1 + e)$$

$$\epsilon = \ln\left(\frac{A_0}{A}\right) = \ln\left(\frac{100}{100 - RA}\right) = \ln(1 + e) \qquad (5.5)$$

where RA is the percent reduction in area of the specimen.

Representative "engineering" and "true" stress–strain behavior is shown in Fig. 5.1. The elastic region has also been plotted on an extended scale to better indicate the yield strength and modulus of elasticity. The following terms from monotonic tensile tests have been and still are included in fatigue design:

S_y = yield strength, MPa (ksi)
S_u = ultimate strength, MPa (ksi)
σ_f = true fracture strength, MPa (ksi)
ϵ_f = true fracture strain or ductility = $\ln[100/(100 - RA)]$
n = strain hardening exponent from Eq. 3.2
K = strength coefficient from Eq. 3.2, MPa (ksi)

The subscript f represents fracture. The true fracture strength is usually corrected for necking, which causes a biaxial state of stress at the neck

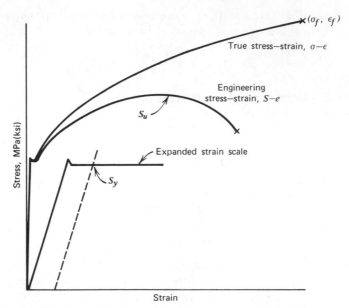

FIGURE 5.1   Engineering and true stress–strain behavior.

surface and a triaxial state of stress at the neck interior. The Bridgman correction factor is used to compensate for this triaxial state of stress [2]. Representative values of these monotonic tensile material properties of unnotched specimens, along with accompanying fatigue properties, are given in Tables A.1 and A.2 for selected engineering alloys.

Monotonic tensile stress–strain behavior of notched specimens may be similar or quite different from that of unnotched specimens. For most notched tensile tests, the ductility is reduced. In low and intermediate strength ductile materials, the ultimate strength is usually increased in notched specimens, while for high strength brittle materials the ultimate strength can be reduced in notched specimens. Notch strength is commonly unavailable in material property handbooks; however, the above differences do exist.

5.2   FATIGUE SPECTRA, TEST MACHINES, AND SPECIMENS

5.2.1   Fatigue Histories or Spectra

Components and structures are subjected to quite diverse load histories. In some cases they may be rather simple and repetitive, while at the other extreme they may be completely random. The randomness, however, may contain substantial portions of more deterministic loading. For example, the ground–air–ground cycle of an aircraft has substantial similarity from flight to flight. A schematić load history of one such flight is shown in Fig. 5.2*a*. The history consists primarily of taxi, flight with gusts, landing, and taxi loads. Standardized test histories have been developed to aid in comparative test programs for transport aircraft [3]. Three typical histories obtained by the SAE Fatigue Design and Evaluation Committee from actual ground vehicle components are shown in Fig. 5.2*b* [4]. These histories were also used in comparative test programs to better establish fatigue life prediction techniques [4]. A typical load history of short-span bridges that has been used in bridge fatigue life studies is shown in Fig. 5.2*c* [5]. These five load histories are typical of those found in real-life engineering situations. If closed-loop electrohydraulic test systems are used, real-life loading histories can be applied directly to small test specimens, components, subassemblies, and even entire products. Historically, complex load histories were often, and still are, replaced in test programs by more simplified loadings, such as the block programs shown in Figs. 5.2*d* and 5.2*e*, and by constant amplitude tests shown in Fig. 5.3.

Nomenclature used in fatigue design has been superimposed on the constant amplitude stress versus time curve in Fig. 5.3. Definitions of alternating, mean, maximum, minimum, and range of stress (strain or stress intensity factor) are indicated. The algebraic relationships among these terms are:

$$S_a = \frac{S_{max} - S_{min}}{2}$$

$$S_m = \frac{S_{max} + S_{min}}{2}$$

$$S_{max} = S_m + S_a \qquad (5.6)$$

$$S_{min} = S_m - S_a$$

The stress range is just twice the alternating stress. Tensile or compressive stresses or strains are taken algebraically as positive and negative, respectively. Alternating stress or strain is an absolute value. The two common stress ratios R and A used frequently in the fatigue literature are:

$$R = \frac{S_{min}}{S_{max}}$$

$$A = \frac{S_a}{S_m} \qquad (5.7)$$

Stresses in Eqs. 5.6 and 5.7 can be replaced with strain or stress intensity factors. One cycle is the smallest segment of the stress or strain versus time history which is repeated periodically, as shown in Fig. 5.3. Under variable amplitude loading, the definition of one cycle is not clear and hence reversals of stress or strain are often considered. In constant amplitude loading,

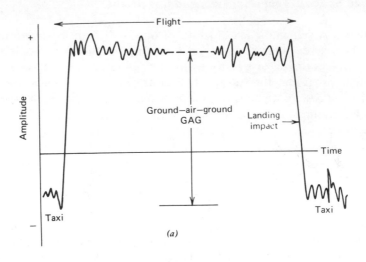

(a)

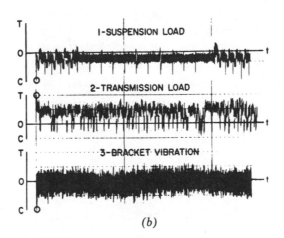

(b)

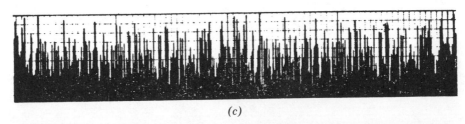

(c)

FIGURE 5.2 Real and block program load spectra [4, 5] (reprinted by permission of the Society of Automotive Engineers and the American Society for Testing and Materials). (a) Schematic spectrum of flight loads. (b) SAE representative test spectra [4]. (c) Load spectrum from the National Cooperative Highway Research Program [5]. (d) Programmed 6 load level test. (e) Random block program loading.

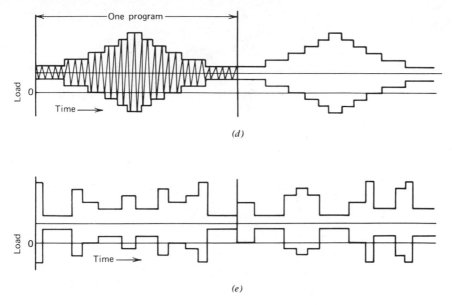

(d)

(e)

FIGURE 5.2 *(Continued)*

one cycle contains two reversals, while in variable amplitude loading a defined cycle may contain many reversals.

5.2.2 Fatigue Test Machines

Systematic, constant amplitude fatigue testing was first initiated by Wöhler in the 1850s on railway axles. Figure 5.4 schematically shows several common constant amplitude fatigue test machines. Rotating bending machines are shown in Figs. 5.4a and 5.4b. The test machine in Fig. 5.4b produces a uniform, pure bending moment over the entire test section of the speci-

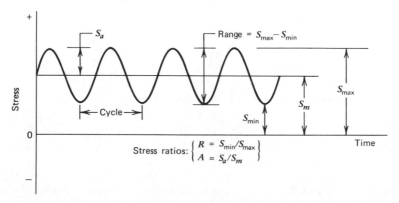

FIGURE 5.3 Nomenclature for constant amplitude cyclic loading.

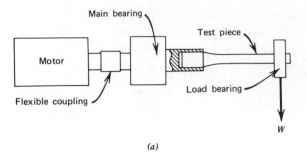

(a)

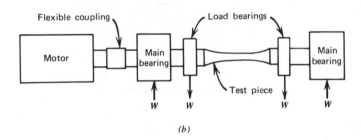

(b)

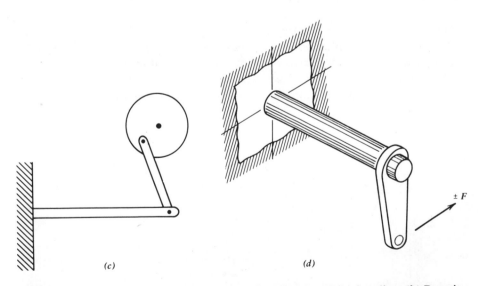

(c)   (d)

FIGURE 5.4 Fatigue testing machines. (a) Cantilever rotating bending. (b) Rotating pure bending. (c) Bending cantilever eccentric crank. (d) Combined torsion and bending.

men. These test machines are known as constant load amplitude machines
because, despite changes in material properties or crack growth, the load
amplitudes do not change. A constant deflection amplitude cantilever bending test machine is shown schematically in Fig. 5.4c. Since the rotating
eccentric crank produces a constant deflection amplitude, the load amplitude changes with specimen cyclic hardening or softening and decreases as
cracks in the specimen initiate and propagate. For a given initial stress
amplitude the constant deflection test machines normally give longer fatigue
life, because of the decrease in load amplitude. The eccentric crank test

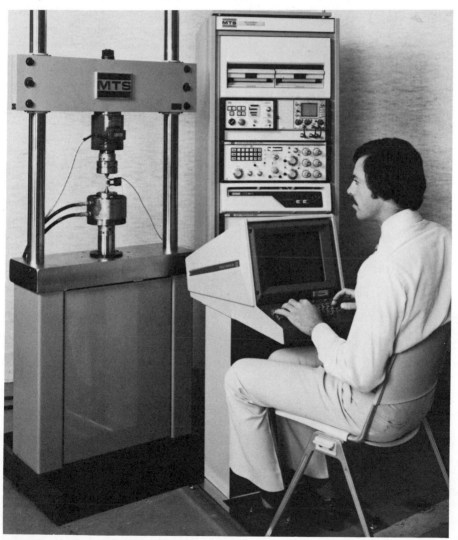

FIGURE 5.5   Closed-loop electrohydraulic test system utilizing its own minicomputer (courtesy of MTS Corp.).

machines, however, do have an advantage over the rotating bending test machines in that the mean deflection, and hence the initial mean stress, can be varied. A common test setup for in-phase combined torsion and bending is shown in Fig. 5.4d. Many additional test machines have been designed over the years, and an important recent contribution to fatigue testing has been the closed-loop electrohydraulic test system. A modern test system utilizing its own minicomputer is shown in Fig. 5.5. These test systems can perform constant or variable amplitude load, strain, or deformation controlled tests on small specimens or can be utilized with hydraulic jacks for components, subassemblies, or whole structures. Test methods and procedures for the above fatigue machines can be found in references 6 through 8.

5.2.3 Fatigue Test Specimens

Common test specimens for obtaining fatigue data are shown in Fig. 5.6. Specimens *a* through *f* have been used to obtain total fatigue life, including

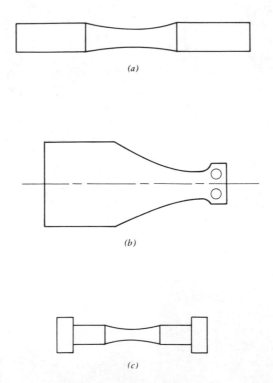

(a)

(b)

(c)

FIGURE 5.6   Fatigue specimens: (*a*) rotating bending, (*b*) cantilever flat sheet, (*c*) axial hour glass, (*d*) axial, (*e*) torsion, (*f*) combined stress, (*g*) axial cracked sheet, (*h*) part-through crack, (*i*) compact specimen, (*j*) three-point bend specimen.

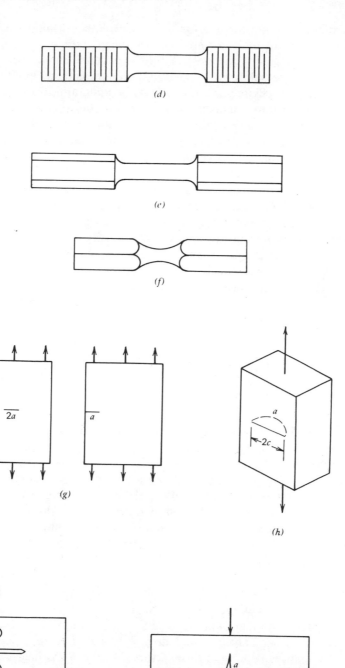

FIGURE 5.6 (*Continued*)

crack initiation and propagation. These specimens usually have finely polished surfaces to minimize surface roughness effects. No distinction between crack initiation and propagation is normally made with these specimens. Bending, axial, torsion, and combined torsion and bending specimens are included in specimens a through f. Bending and axial specimens with circular cross sections usually have diameters between about 3 and 10 mm ($\frac{1}{8}$ to $\frac{3}{8}$ in.). Stress concentration influence can be studied with these specimens by machining in grooves, notches, or holes. Recommended surface preparations for these fatigue specimens can be found in references 6 and 8.

Specimens g through j in Fig. 5.6 have been used for obtaining fatigue crack growth data. In all cases a thin slit or groove with a very small root radius is machined into the specimen. A small pretest fatigue crack is then formed at this root radius by low stress intensity range cyclic loading. After this sharp pretest fatigue crack has been formed, the real fatigue test can begin. Recommended specimen preparation, test procedure, and data reduction have been formulated for crack growth rate tests above 10^{-8} m/cycle (4×10^{-7} in./cycle) using center cracked sheets (specimen g) or compact specimens (specimen i) [9].

5.3   STRESS–LIFE CURVES, S–N

5.3.1   General Behavior

Two typical S–N curves obtained under load or stress control test conditions with smooth specimens are shown in Fig. 5.7. Here S is the applied stress, usually taken as the alternating stress S_a, and N is the number of cycles or life to failure, where failure is defined as fracture. Constant amplitude S–N curves of this type are plotted on semilog or log–log coordinates and often contain fewer data points than shown. Figure 5.7 shows typical scatter. Figure 5.7a shows a continuous sloping curve, while Fig. 5.7b shows a discontinuity or "knee" in the S–N curve. This knee has been found in only a few materials (notably the low strength steels) between 10^6 and 10^7 cycles under noncorrosive conditions. Most materials do not contain the "knee" even under carefully controlled environmental conditions. Under corrosive environments all S–N data invariably have a continuous sloping curve. When sufficient data are taken at several stress levels, S–N curves are usually drawn through median points and thus represent 50 percent expected failures. Common terms used with the S–N diagram are "fatigue life," "fatigue strength," and "fatigue limit." ASTM definitions of these terms are given below [10].

The fatigue life, N, is the number of cycles of stress or strain of a specified character that a given specimen sustains before failure of a specified nature occurs. Fatigue strength is a hypothetical value of stress at failure for

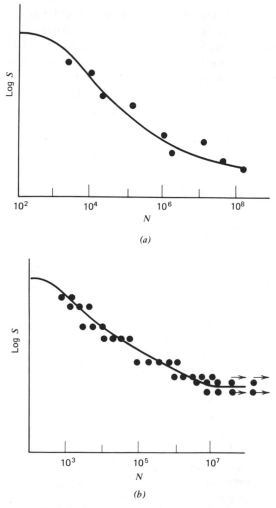

(a)

(b)

FIGURE 5.7   Typical S–N diagrams.

exactly N cycles as determined from an S-N diagram. Fatigue limit, S_f, is the limiting value of the median fatigue strength as N becomes very large. The above definitions are all based on median values or 50 percent survival.

In Chapter 3 it is emphasized that fatigue consists of crack initiation, propagation, and final fracture. Figure 5.7 does not separate crack initiation from propagation, and only the total life to fracture is given. Let us assume a reasonable crack initiation life defined by a crack length of say 0.25 mm (0.01 in.). This dimension, which is easily seen at 10× to 20× magnification, can relate to engineering dimensions and can represent a small macrocrack. The number of cycles to form this small crack in smooth unnotched or notched fatigue specimens and components can range from a few percent to almost the entire life. The size of the final crack at fracture depends on the fracture toughness of the material and the stress level. The higher stress levels have shorter critical crack sizes and the lower stress levels have larger critical crack sizes. Assuming test specimens are between 3 and 10 mm ($\frac{1}{8}$ and $\frac{3}{8}$ in.) in diameter and that critical crack lengths can vary from 10 to 70 percent of the diameter, then critical crack lengths for these small test specimens vary from about 0.25 to 6 mm (0.01 to 0.25 in.). These dimensions have substantial importance in later chapters on fatigue life prediction of components.

5.3.2 Fatigue Limit under Fully Reversed Uniaxial Loading

The fatigue limit has historically been a prime consideration for long-life fatigue design. For a given material the fatigue limit has an enormous range depending on surface finish, size, type of loading, temperature, corrosive environment, mean and self-stresses (residual stresses), and stress concentrations. Considering the fatigue limit based on a nominal alternating stress, S_a, this value has ranged from essentially 1 to 70 percent of the ultimate tensile strength. An example of a case where the fatigue limit may be approximately 1 percent of S_u is a high strength steel with a sharp notch subjected to a high mean tensile stress in a very corrosive atmosphere. An example of a case when the fatigue limit might approach 70 percent of S_u is a notched medium strength steel in an inert atmosphere with appreciable compressive self-stresses (residual stresses) in the notch region.

Most long-life S-N fatigue data available consist of fully reversed ($S_m = 0$) uniaxial fatigue strengths or fatigue limits of small highly polished unnotched specimens based on 10^6 to 5×10^8 cycles to failure in laboratory air environment. Representative static tensile properties and fatigue limits of metals obtained under the above conditions are given in Table A.1. Most of these values are fairly independent of test frequency between about 5 and 100 Hz in noncorrosive environments. The data for the aluminum alloys were originally stamped ''NOT FOR DESIGN'' by the Aluminum Association [11]. This apparently was due to product liability concern, and also to the fact that these data were obtained from small rotating bending unnotched conditions, not from components subjected to actual service or field conditions and environments. The fatigue limits given in Table A.1 *must* be substantially reduced in most cases before they can be used in design situations [12–14]. For example, 15 to 25 percent reduction in these values for just size effect alone is not unreasonable for specimens greater than 12.5 mm (0.5 in.) [14].

A common procedure to partially compare materials is to plot the unnotched fully reversed fatigue limit, S_f, obtained under similar ideal laboratory conditions described above, versus the ultimate tensile strength, S_u. The ratio of S_f/S_u is called the fatigue ratio. Figure 5.8 shows plots for many tests of steels, irons, aluminum alloys, and copper alloys [14]. Lines of constant fatigue ratio, S_f/S_u, equal to 0.35, 0.5, and 0.6 are superimposed on the data. It is quite clear that S_f/S_u varies from about 0.2 to 0.65 for this data. There is a tendency to generalize that S_f increases linearly with S_u. A careful examination of Fig. 5.8 shows this is quite incorrect and that data bands tend to bend over at the higher ultimate strengths. This tendency, however, can be overcome by suitable surface treatment as explained in Chapters 6 and 7.

Figure 5.8a for steels indicates substantial data are clustered near the fatigue ratio $S_f/S_u = 0.5$ for the low and medium strength steels. The data, however, actually fall between 0.35 and 0.6 for $S_u \leq 1400$ Mpa (200 ksi). For $S_u > 1400$ MPa (200 ksi), S_f does not increase significantly. Thus very common loosely used criteria for unnotched, highly polished, small bending specimen fatigue limits for steels are:

$$S_f \approx 0.5 \, S_u \qquad \text{for } S_u \leq 1400 \text{ MPa (200 ksi)} \qquad (5.8a)$$

$$S_f \approx 700 \text{ MPa (100 ksi)} \qquad \text{for } S_u \geq 1400 \text{ MPa (200 ksi)} \qquad (5.8b)$$

By suitable surface treatments, however, the fatigue limits of the high strength steels can also be increased to agree more with Eq. 5.8a. Equations 5.8 are not unreasonable for the small highly polished specimens tested, but empirical reduction factors for surface finish, size, stress concentration, temperature, and corrosion must also be considered [12–14]. We strongly warn against using a design fatigue limit equal to one-half the ultimate strength for steels. Most data in Fig. 5.8 for irons, aluminum, and copper alloys fall below the 0.5 fatigue ratio. The aluminum and copper alloy data bands bend over at higher strengths as do the bands for the steels. Thus high strength steels, aluminum, and copper alloys do not exhibit corresponding high unnotched fatigue limits.

Data for steels similar to those in Fig. 5.8a can be generalized into schematic scatter bands, as shown in Fig. 5.9 [15]. Here it is seen that severely notched and/or corroding specimens have substantially lower fatigue limits than unnotched specimens. But again, this can be significantly altered in many cases by proper surface operations described in Chapter 6 and 7.

5.3.3   Mean Stress Effects

The alternating stress, S_a, and the mean stress, S_m, are defined in Fig. 5.3 and Eq. 5.6. The mean stress can have a substantial influence on fatigue behavior as is shown in Fig. 5.10, where S_a is plotted against the number of cycles to failure N. It is seen that, in general, tensile mean stresses are detrimental and compressive mean stresses are beneficial. At intermediate or high stress levels under load control test conditions, substantial cyclic creep, which increases the mean strain, can occur in the presence of mean stresses, as shown by Landgraf [16] in Fig. 5.11.

Substantial investigation of tensile mean stress influence on long-life fatigue strength has been made. Typical dimensionless plots are shown in Fig.

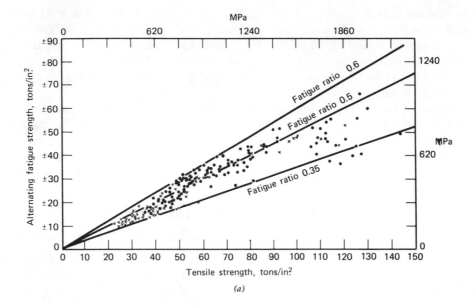

(a)

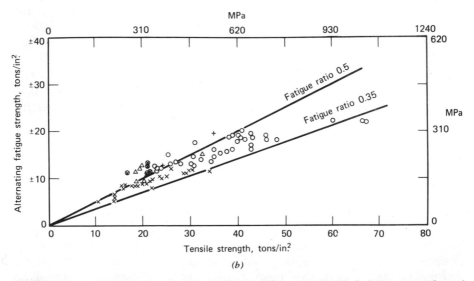

(b)

FIGURE 5.8 Relation between rotating bending unnotched fatigue strength and
ultimate tensile strength [14] (reprinted with permission of Pergamon Press Ltd.).
(*a*) Carbon and alloy steels (10^7 to 10^8 cycles): (●) alloy steels, (×) carbon steels.
(*b*) Wrought and cast irons (10^7 cycles): (×) flake-graphite cast iron, (○) nodular
cast iron, (+) malleable cast iron, (△) ingot iron, (⊗) wrought iron. (*c*) Aluminum
alloys based on 10^8 cycles: (×) wrought, (⊗) cast. (*d*) Wrought copper alloys.

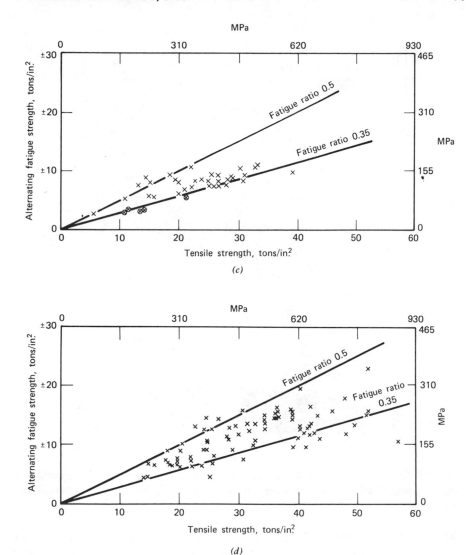

FIGURE 5.8 (*Continued*)

5.12 for steels and aluminum alloys, where S_a/S_f versus S_m/S_u is plotted. S_f is the fully reversed, ($S_m = 0$, $R = -1$), fatigue strength and S_u is the ultimate tensile strength. Similar behavior exists for other alloys. Substantial scatter exists, but the general trend indicating that tensile mean stresses are detrimental is quite evident. Small tensile mean stresses, however, often have only a small effect. It appears that many of the data fall between the straight and curved lines. The straight line is the modified Goodman line and the curve is the Gerber parabola. Two additional relationships have been formulated by replacing S_u with S_y (Soderberg line) and S_u with σ_f

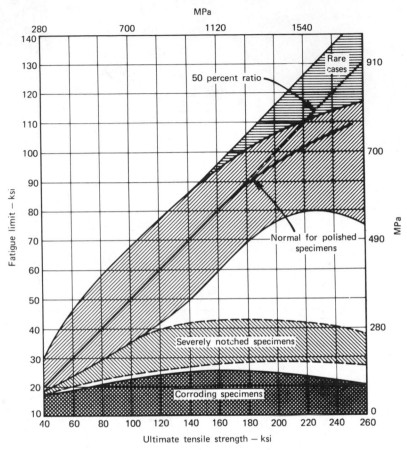

FIGURE 5.9 General relationship between fatigue limits and tensile strength for polished unnotched and for corroding steel specimens [15] (reprinted with permission of John Wiley and Sons, Inc.).

(Morrow line) where S_y and σ_f are the tensile yield strength and true fracture stress, respectively. The following equations represent these tensile mean stress effects:

Modified Goodman, $$\frac{S_a}{S_f} + \frac{S_m}{S_u} = 1 \qquad (5.9a)$$

Gerber, $$\frac{S_a}{S_f} + \left(\frac{S_m}{S_u}\right)^2 = 1 \qquad (5.9b)$$

Soderberg, $$\frac{S_a}{S_f} + \frac{S_m}{S_y} = 1 \qquad (5.9c)$$

Morrow, $$\frac{S_a}{S_f} + \frac{S_m}{\sigma_f} = 1 \qquad (5.9d)$$

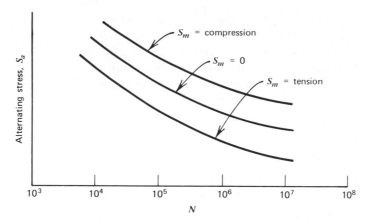

FIGURE 5.10 Effect of mean stress on fatigue life.

All four expressions have been used in fatigue design when modified for notches, size, surface finish, and environmental effects. A yield criterion has been used in conjunction with these expressions.

Figure 5.13 shows a typical constant life fatigue diagram that incorporates median fatigue life, tensile mean stress, and notch effects basd on S-N data similar to those in Figs. 5.10 and 5.12 [17]. Many such diagrams are available to design engineers in *Mil Handbook 5* for metals applicable to the aerospace field [18]. These figures readily provide maximum, minimum, mean, and alternating nominal stresses for a given life of small test specimens.

Figures 5.12 and 5.13 do not provide information on compressive mean stress effects, as shown by Sines [19] in Fig. 5.14 for several steels and

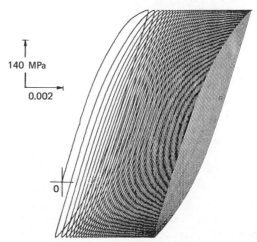

FIGURE 5.11 Cyclic creep under load control constant amplitude testing, $S_m > 0$ [16] (reprinted by permission of the American Society for Testing and Materials).

aluminum alloys. It is seen that these compressive mean stresses cause increases of up to 50 percent in the alternating fatigue strength. Even higher increases have been shown [14]. This increase is too often overlooked, since compressive self-stresses (residual stresses) can cause similar beneficial behavior. Fuchs [20] developed general quantitative criteria for long-life

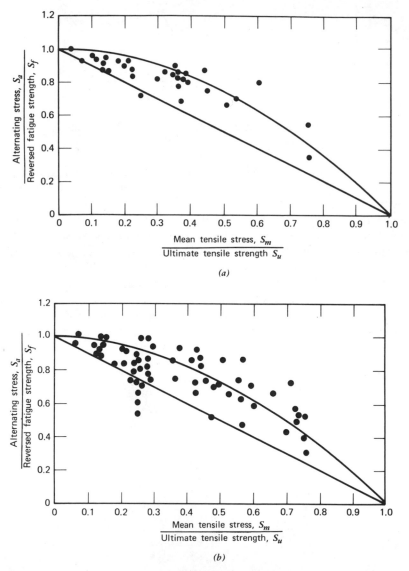

(a)

(b)

FIGURE 5.12 Effect of mean stress on alternating fatigue strength at long life [14] (reprinted with permission of Pergamon Press Ltd.). (*a*) Steels based on ~10^7 cycles. (*b*) Aluminum alloys based on ~5×10^7 cycles.

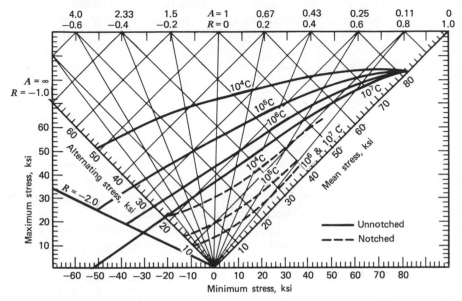

FIGURE 5.13 Constant-life fatigue diagram for (——) unnotched and (- - -) notched 7075-T6 aluminum alloy specimens [17]. S_u = 82 ksi 2000 cpm; K_t = 3.4.

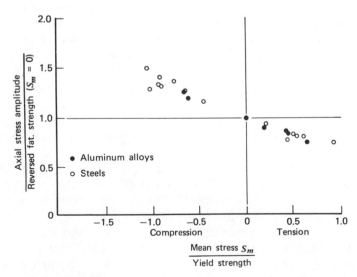

FIGURE 5.14 Compressive and tensile mean stress effect [19]. (●) Aluminum alloys; (○) steels.

fatigue design involving both tensile and compressive mean or self-stresses, notches, and multiaxial stresses. These are covered in later chapters.

5.4 LOW CYCLE FATIGUE, ϵ–N

In a notched component or specimen subjected to cyclic external loads, the behavior of material at the root of the notch is best considered in terms of strain. As long as there is elastic constraint surrounding a local plastic zone at the notch the strains can be calculated more easily than the stress. This concept has motivated a finite fatigue life design philosophy based on re-lating the fatigue life of notched parts to the life of small unnotched speci-mens that are cycled to the same strains as the material at the notch root. This is called strain control. Reasonable expected fatigue life, based on the initiation or formation of small macrocracks can then be determined know-ing the local strain–time history at a notch in a component and the un-notched strain–life fatigue properties of the material and assuming a rea-sonable cumulative damage theory. The SAE Fatigue Design and Evaluation Committee has recently completed an extensive interlaboratory test and evaluation program that included evaluating this technique [4]. The results were very encouraging. The remaining fatigue crack growth life of a com-ponent can be analyzed using fracture mechanics concepts.

Substantial strain–life fatigue data needed for the above procedure have been accumulated and published in the form of simplified fatigue material properties [21]. Some of these data are included in Table A.2 for selected engineering alloys. These properties are obtained from small, polished, unnotched axial fatigue specimens similar to those in Figs. 5.6c and 5.6d. Tests are performed under constant amplitude fully reversed cycles of strain, as shown in Fig. 3.10. Steady-state hysteresis loops can predominate through most of the fatigue life, and these can be reduced to elastic and plastic strain ranges or amplitudes. Cycles to failure can involve from about 10 to 10^6 cycles and frequencies range from about 0.2 to 5 Hz. Beyond 10^6 cycles, load or stress controlled higher frequency tests can be run, because of the small plastic strains and the greater time to failure. The strain–life curves are often called low cycle fatigue data because much of the data are for less than about 10^5 cycles.

Strain–life fatigue curves plotted on log–log scales are shown schemati-cally in Fig. 5.15, where N or $2N$ is the number of cycles or reversals to failure, respectively. Failure criteria for strain–life curves have not been consistently defined in that failure may be the life to a small detectable crack, life to a certain percentage decrease in load amplitude, or life to fracture. Differences in fatigue life depending on these three criteria may be small or appreciable. Crack lengths at these failure criteria are discussed later in this section.

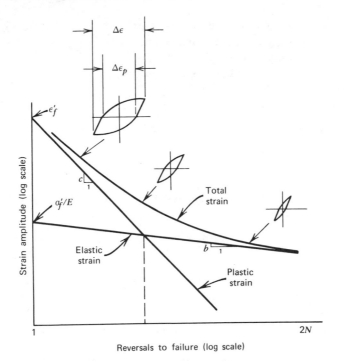

FIGURE 5.15 Strain–life curves showing total elastic and plastic strain components.

The total strain amplitude in Fig. 5.15 has been resolved into elastic and plastic strain components from the steady-state hysteresis loops. At a given life N, the total strain is the sum of the elastic and plastic strains. Both the elastic and plastic curves can be approximated as straight lines. At large strains or short lives, the plastic strain component is predominant, and at small strains or longer lives the elastic strain component is predominant. This is indicated by the straight line curves and the hysteresis loop sizes in Fig. 5.15. The intercepts of the two straight lines at $2N = 1$ are ϵ_f' for the plastic component and σ_f'/E for the elastic component. The slopes are c and b, respectively. This provides the following equation for strain–life data of small smooth axial specimens:

$$\frac{\Delta\epsilon}{2} = \frac{\Delta\epsilon_e}{2} + \frac{\Delta\epsilon_p}{2}$$

$$= \frac{\sigma_f'}{E}(2N)^b + \epsilon_f' \, (2N)^c$$

(5.10)

where $\Delta\epsilon/2$ = total strain amplitude
 $\Delta\epsilon_e/2$ = elastic strain amplitude = $\Delta\sigma/2E = \sigma_a/E$
 $\Delta\epsilon_p/2$ = plastic strain amplitude = $\Delta\epsilon/2 - \Delta\epsilon_e/2$

$$\epsilon_f' = \text{fatigue ductility coefficient}$$
$$c = \text{fatigue ductility exponent}$$
$$\sigma_f' = \text{fatigue strength coefficient}$$
$$b = \text{fatigue strength exponent}$$
$$E = \text{modulus of elasticity}$$
$$\Delta\sigma/2 = \sigma_a = \text{stress amplitude}$$

The straight line elastic behavior can be transformed to

$$\frac{\Delta\sigma}{2} = \sigma_a = \sigma_f' \, (2N)^b \tag{5.11}$$

which is Basquin's equation proposed in 1910 [22]. The relation between plastic strain and life is

$$\frac{\Delta\epsilon_p}{2} = \epsilon_f' \, (2N)^c \tag{5.12}$$

which is the Manson-Coffin relationship first proposed in the early 1960s [23, 24]. The exponent c ranges from about -0.5 to -0.7, with -0.6 as a representative value. The exponent b ranges from about -0.06 to -0.14, with -0.1 as a representative value. The term ϵ_f' is somewhat related to the true fracture strain, ϵ_f, in a monotonic tensile test and in most cases ranges from about 0.35 to 1.0 times ϵ_f. The coefficient σ_f' is somewhat related to the true fracture stress, σ_f, in a monotonic tensile test. Sometimes ϵ_f' and σ_f' may be taken as ϵ_f and σ_f, respectively, as a rough first approximation. A typical complete strain–life curve with data points is shown in Fig. 5.16 for 4340 steel [25]. Eleven test specimens were used to form these strain–life curves.

Manson [26] has simplified Eq. 5.10 even further with his method of universal slopes where

$$\Delta\epsilon = 3.5 \frac{S_u}{E} \, (N)^{-0.12} + \epsilon_f^{0.6} \, (N)^{-0.6} \tag{5.13}$$

S_u, E, and ϵ_f are all obtained from a monotonic tensile test. He assumes the two exponents are fixed for all materials and only S_u, E, and ϵ_f control the fatigue behavior. Figure 5.17 indicates good agreement between un-notched smooth specimen experimental data and the universal slopes method. Thus the universal slopes method can be a first approximation for the fully reversed strain–life curve for unnotched smooth specimens, based on monotonic tensile properties.

The general differences of metals under strain-controlled tests are shown in Fig. 5.18. Many materials have about the same life at a total strain amplitude of 0.01. At larger strains, increased life of unnotched smooth test specimens is dependent more on ductility, while at smaller strains better life is obtained from stronger materials. Life here refers to the initiation of a small detectable crack, a percentage decrease in the load amplitude that is caused by crack initiation and growth, or final fracture. For the final

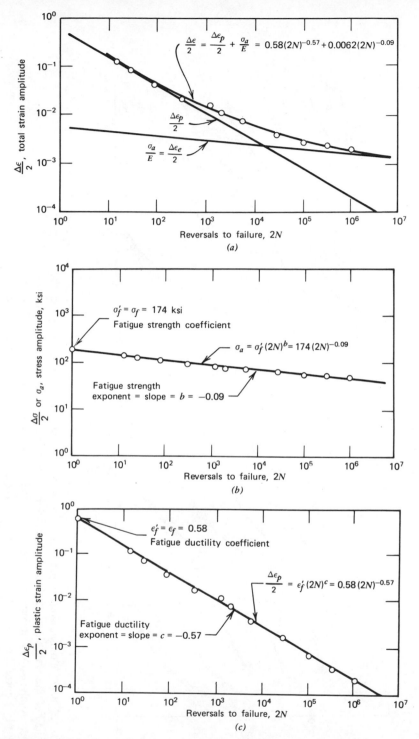

FIGURE 5.16 Low cycle fatigue behavior of annealed 4340 steel [25] (reprinted with permission of the Society of Automotive Engineers). (*a*) Total strain amplitude. (*b*) Elastic strain amplitude $\times E$. (*c*) Plastic strain amplitude.

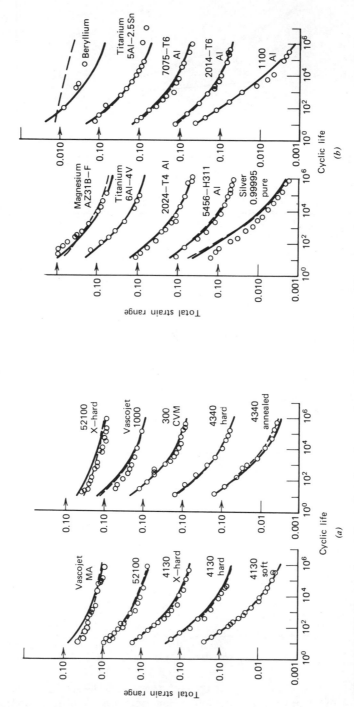

FIGURE 5.17 Comparison of predicted and experimental axial fatigue life using the method of universal slopes [26] (reprinted with permission of the Society for Experimental Stress Analysis). (*a*) Steels. (*b*) Nonferrous metals. (○) Experimental; (- -) predicted by four-point correlation; (——) predicted by universal slopes. Note shift in strain scales.

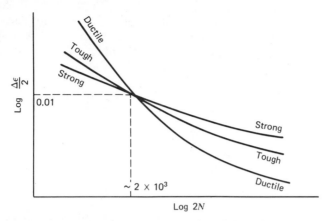

FIGURE 5.18 Schematic of strain–life curves for different materials.

fracture criterion, the crack would grow to about 10 to 70 percent of the
test specimen cross section. Since strain–life test specimens are usually
between about 3 and 6 mm ($\frac{1}{8}$ and $\frac{1}{4}$ in.), this implies the strain–life fracture
criteria are based on cracks growing to a depth of about 0.25 to 5 mm (0.01
to 0.2 in.). The actual value depends on the strain amplitude, modulus of
elasticity, and the material's fracture toughness. The other two criteria are
based on life to cracks, which would be smaller than those at fracture. In
general, cracks less than 0.25 mm (0.01 in.) would not be readily observed
in these tests and would probably not cause sufficient decrease in the load
amplitude to terminate a test. Thus a reasonable important conclusion concerning failure criteria in strain–life testing of unnotched smooth specimens
is that the life to failure means life to fatigue crack lengths of between 0.25
to 5 mm (0.01 to 0.2 in.). This criterion is dependent on strain amplitude,
modulus of elasticity, and fracture toughness.

Low cycle strain–life fatigue data in Table A.2 and in reference 21 were
obtained under the above conditions. Both monotonic tensile properties and
strain–life material properties are included for completeness in both SI and
American/British units. Terms in Table A.2 are described earlier. Data for
the last two columns, S_f and S_f/S_u, were obtained by substituting the
proper material constants and $2N = 10^7$ cycles into Eq. 5.10 to obtain S_f.
This value is an approximation of reasonable long-life unnotched smooth
axial fatigue strengths and should be somewhat compatible with values of
S_f in Table A.1. Values in Table A.2 should be about 10 to 25 percent lower
than those in Table A.1 for a given material, because of the difference
between bending and axial long-life fatigue strengths. The fatigue ratio, $S_f/$
S_u (last column), does have values lower than those given in Fig. 5.8.
Material properties in Table A.2 also omit influences of surface finish, size,
stress concentration, temperature, and corrosion. These must be included
in fatigue design, and thus values in Table A.2 do not represent final fatigue
design properties.

The inclusion of mean stress or mean strain effects in fatigue life prediction methods involving strain-life data is very complex. One method is to replace σ_f' with $\sigma_f' - \sigma_m$ in Eq. 5.10 [25], where σ_m is the mean stress, such that

$$\frac{\Delta\epsilon}{2} = \frac{(\sigma_f' - \sigma_m)(2N)^b}{E} + \epsilon_f'(2N)^c \tag{5.14}$$

here σ_m is taken positive for tensile values and negative for compressive values. Another equation suggested by Smith et al. [27], based on strain-life test data at fracture obtained with various mean stresses, is

$$\sigma_{max}\epsilon_a E = (\sigma_f')^2(2N)^{2b} + \sigma_f'\epsilon_f' E(2N)^{b+c} \tag{5.15}$$

where $\sigma_{max} = \sigma_m + \sigma_a$ and ϵ_a is the alternating strain. If σ_{max} is zero, Eq. 5.15 predicts infinite life, which implies that tension must be present for fatigue fractures to occur. Both Eqs 5.14 and 5.15 have been used to handle mean stress effects [4, 25].

5.5 FATIGUE CRACK GROWTH, *da/dN–ΔK*

Linear elastic fracture mechanics (LEFM) concepts are most useful to correlate fatigue crack growth behavior, as is shown in Chapter 4. The form of this correlation for constant amplitude loading is usually a log–log plot of fatigue crack growth rate, da/dN, in m/cycle (in./cycle), versus the opening mode stress intensity factor range ΔK_I (or ΔK), in MPa$\sqrt{m}$ (ksi$\sqrt{in}$.), where ΔK_I is defined as

$$\begin{aligned}\Delta K_I = \Delta K &= K_{max} - K_{min} \\ &= S_{max}\sqrt{\pi a}\,\alpha - S_{min}\sqrt{\pi a}\,\alpha \end{aligned} \tag{5.16}$$

Since the stress intensity factor $K_I = S\sqrt{\pi a}\,\alpha$ is undefined in compression, K_{min} is taken as zero if S_{min} is compression.

The elastic stress intensity factor is applicable to fatigue crack growth even in low strength, high ductility materials because K_I values needed to cause fatigue crack growth are quite low. Thus plastic zone sizes at the crack tip are often small enough to allow an LEFM approach. At very high crack growth rates some difficulties can occur, as a result of large plastic zone sizes, but this is often not a problem because very little fatigue life may be involved.

5.5.1 Sigmoidal Shaped *da/dN–ΔK* Curve

Many fatigue crack growth data have been obtained under constant load amplitude test conditions using sharp notched fatigue cracked specimens similar to those in Figs. 5.6*g* through 5.6*j*. Mode I fatigue crack growth has received the greatest attention, because this is the predominant mode of macroscopic fatigue crack growth. K_{II} and K_{III} usually have only second order effects on both crack direction and crack growth rates. Crack growth is usually measured with optical, compliance, ultrasonic, eddy current, electrical potential, or accoustic emission techniques. Figure 4.1 shows typical crack length versus applied cycles for constant amplitude tests. Crack growth rates, da/dN, as a function of ΔK_I, can be obtained at consecutive positions along these curves using graphical or numerical methods [9]. A typical complete log–log plot of da/dN versus ΔK_I is shown schematically in Figs. 4.2 and 5.19. This curve has a sigmoidal shape that can be divided into three major regions, as shown in Fig. 5.19. Region I indicates a threshold value ΔK_{th}, below which there is no observable crack growth. This threshold occurs at crack growth rates on the order of 2.5×10^{-10} m/cycle (10^{-8} in./cycle) or less. Below ΔK_{th}, fatigue cracks behave as nonpropagating cracks. Region II shows essentially a linear relationship between log da/dN and log ΔK_I, which corresponds to the formula

$$\frac{da}{dN} = A(\Delta K_I)^n \tag{5.17}$$

first suggested by Paris [28]. Here n is the slope of the curve and A is the coefficient found by extending the straight line to $\Delta K_I = 1$ MPa$\sqrt{m}$ (ksi$\sqrt{in.}$). In region III the crack growth rates are very high and little fatigue

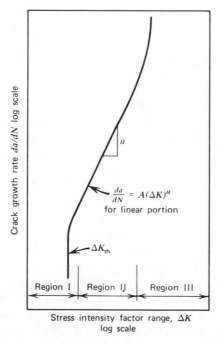

FIGURE 5.19 Schematic sigmoidal behavior of fatigue crack growth rate versus ΔK.

crack growth life is involved. Region III may have the least importance in most fatigue situations. This region is controlled primarily by fracture toughness K_c or K_{Ic}.

The fatigue crack growth behavior shown in Fig. 5.19 is essentially the same for different specimens or components taken from a given material, because the stress intensity factor range is the principal controlling factor in fatigue crack growth. This allows fatigue crack growth rate versus ΔK_I data obtained under constant amplitude conditions with simple specimen configurations to be used in design situations. Knowing the stress intensity factor expression, K_I, for a given component and loading, the fatigue crack growth life of the component can be obtained by integrating Eq. 5.17 between limits of initial crack size and final crack size. The greatest usage of Fig. 5.19 data has been in fail-safe design of aircraft and nuclear energy systems along with fractographic failure analysis. Crack growth rate behavior has also become important in material selection and comparative prototype designs. However, as with all constant amplitude material fatigue properties, these data do not provide information on fatigue crack growth interaction or sequence effects, which are covered in Chapter 10.

5.5.2 da/dN–ΔK for R = 0

Conventional S-N or ϵ-N fatigue behavior is usually referenced to the fully reversed stress or strain conditions ($R = -1$). Fatigue crack growth data, however, are usually referenced to the pulsating tension condition with $R = 0$ or approximately zero. This is based on the concept that during compression loading the crack is closed and hence no stress intensity factor, K, can exist. The compression loads should thus have little influence on constant amplitude fatigue crack growth behavior. In general, this is fairly realistic, but under variable amplitude loading, compression cycles can be important to fatigue crack growth, as is seen in later chapters.

Regions II and III of the sigmoidal curve have received the widest attention. A standard ASTM practice exists for obtaining crack growth experimental data above rates of 10^{-8} m/cycle (4×10^{-7} in./cycle) [9]. Many mathematical equations depicting fatigue crack growth above threshold levels have been formulated. Hoeppner and Krupp [29] list 33 empirical equations. However, the Paris equation seems to be the most popular equation for $R = 0$ loading. Substantial constant amplitude data can be found in the *Damage Tolerant Handbook* for common aircraft alloys [30].

Barsom [31] has evaluated Eq. 5.17 for a wide variety of steels varying in yield strength from 250 to 2070 MPa (36 to 300 ksi). He shows that the fatigue crack growth rate scatter band for a given ΔK_I, with many ferritic–pearlitic steels, varies by a factor of about 2. Partial results from Barsom [31] are shown in Fig. 5.20. He also found a similar scatter band width for martensitic steels, as shown in Fig. 5.21. He suggested that conservative values of the upper boundaries of these scatter bands could be used in design situations if actual data could not be obtained. These suggested

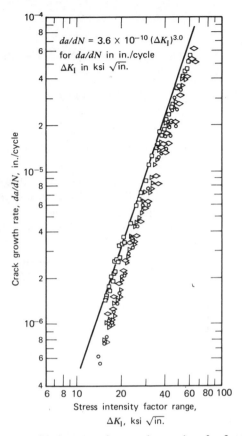

FIGURE 5.20   Summary of fatigue crack growth rate data for ferritic–pearlitic steels [31] (reprinted with permission of the American Society of Mechanical Engineers). (△) A36, (○) ABS-C, (◇) A302-B, (□) A537-A.

conservative equations are:

for ferritic-pearlitic steels

$$\frac{da}{dN} \text{ (m/cycle)} = 6.9 \times 10^{-12}\ (\Delta K\ \text{MPa}\sqrt{m})^{3.0}$$

$$\frac{da}{dN} \text{ (in./cycle)} = 3.6 \times 10^{-10}\ (\Delta K\ \text{ksi}\sqrt{\text{in,}})^{3.0}$$

$$(5.18)$$

for martensitic steels

$$\frac{da}{dN} \text{ (m/cycle)} = 1.35 \times 10^{-10}\ (\Delta K\ \text{MPa}\sqrt{m})^{2.25}$$

$$\frac{da}{dN} \text{ (in./cycle)} = 6.6 \times 10^{-9}\ (\Delta K\ \text{ksi}\sqrt{\text{in.}})^{2.25}$$

$$(5.19)$$

and for austenitic stainless steels

$$\frac{da}{dN} \text{ (m/cycle)} = 5.6 \times 10^{-12} \, (\Delta K \, \text{MPa}\sqrt{\text{m}})^{3.25}$$

$$(5.20)$$

$$\frac{da}{dN} \text{ (in./cycle)} = 3.0 \times 10^{-10} \, (\Delta K \, \text{ksi}\sqrt{\text{in.}})^{3.25}$$

Superposition of these three equations indicates in general that the ferritic–pearlitic steels have better region II fatigue crack growth properties than the martensitic steels or the austenitic stainless steels. The narrowness of the region II scatter bands for the different classifications of steels suggests that large changes in constant amplitude fatigue crack growth life may not be obtained by choosing a slightly different steel. However, this does not take into consideration behavior near or at threshold levels, which does

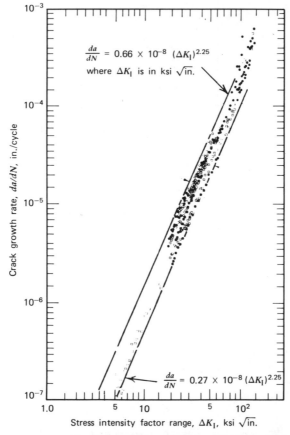

FIGURE 5.21 Summary of fatigue crack growth rate data for martensitic steels [31] (reprinted with permission of the American Society of Mechanical Engineers). (O) 12 Ni steel, (□) 10 Ni steel, (●) Hy-130 steel, (■) HY-80 steel.

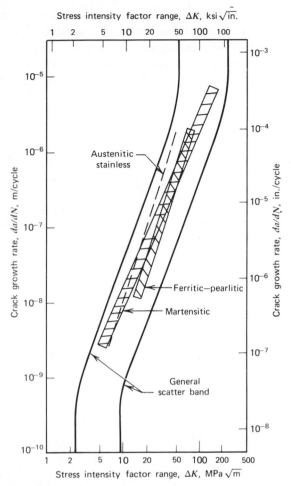

FIGURE 5.22 Superposition of Barsom's scatter bands on the general fatigue crack growth scatter band for steels.

show greater variations as a function of microstructure. An approximate schematic sigmoidal shaped scatter band for steels with Barsom's scatter bands superimposed is shown in Fig. 5.22. It is clear that Barsom's results do not take into consideration complete crack growth rate behavior. Also, despite the numerous tests and materials used to obtain Eqs. 5.18 to 5.20, there are many exceptions. Values of the exponent n have ranged from about 2 to 7. For different aluminum alloys, the fatigue crack growth rates for a given ΔK_I in region II vary much more. The width of the scatter band for different aluminum alloys corresponds to a factor of about 10.

Frequency, wave shape, and thickness effects on constant amplitude fatigue crack growth rates are secondary compared to environmental effects such as corrosion and temperature. They can have influence, but this often is less than that due to different heats or use of different manufacturers. The largest influence comes from variable amplitude loading in various environments. The thickness influence can be greatest in region III because of the inverse relationship between fracture toughness and thickness, as shown in Fig. 4.8, which affects the allowable crack size at fracture. A high fracture toughness is desirable because of the longer final crack size at fracture, which allows easier and less frequent inspection and, therefore, safer components or structures.

Threshold stress intensity factor ranges, ΔK_{th}, are given in Table A.4 for selected engineering alloys. These values are usually less than 9 MPa$\sqrt{m}$ (8 ksi$\sqrt{in.}$) for steels and less than 4 MPa$\sqrt{m}$ (3.5 ksi$\sqrt{in.}$) for aluminum alloys. Values of ΔK_{th} are substantially less than K_{Ic} values given in Fig. 4.9 and Table A.3. In fact, ΔK_{th} can be as low as just several percent of K_{Ic}. The threshold stress intensity factor, ΔK_{th}, has often been considered analogous to the unnotched fatigue limit, S_f, since an applied stress intensity factor range below ΔK_{th} does not cause fatigue crack growth. Figure 5.23

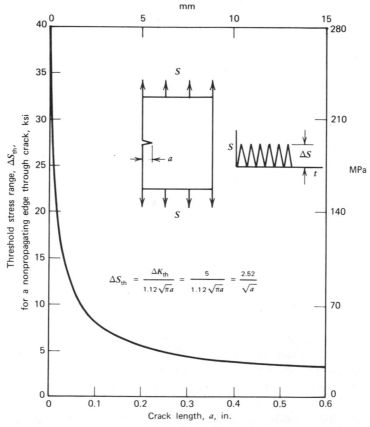

FIGURE 5.23 Threshold stress range for nonpropagating edge through cracks in wide plates, $R = 0$, $\Delta K_{th} = 5.5$ MPa$\sqrt{m}$ (5 ksi$\sqrt{in.}$).

shows the use of ΔK_{th} as a design parameter for no crack growth using a single-edge cracked infinitely wide plate subjected to $R = 0$ loading as shown. In Fig. 5.23, any combination of ΔS and crack length, a, that falls below the curve [assume $\Delta K_{th} = 5.5$ MPa$\sqrt{m}$ (5 ksi$\sqrt{in}$.)] does not cause fatigue crack growth. This criterion comes from setting ΔK_I in Eq. 5.16 equal to ΔK_{th} and using the proper stress intensity factor, K_I, from Fig. 4.5b. It should be noted from Fig. 5.23 that for crack lengths greater than about 2.5 mm (0.1 in.), ΔS_{th} is about 55 MPa (8 ksi) or less. Thus to keep cracks from propagating, the tensile stress must be kept very small. Non-propagating cracks are considered in more detail in later chapters.

5.5.3 Mean Stress Effects

The general influence of mean stress on fatigue crack growth behavior is shown schematically in Fig. 5.24. The stress ratio $R = K_{min}/K_{max} = S_{min}/S_{max}$ is used as the principal parameter. Most mean stress effects on crack growth have been obtained with only tensile stressing, that is, $R \geq 0$. Figure 5.24 indicates that increasing the R ratio (which means increasing both S_{max} and S_{min}) has a tendency to increase the crack growth rates in all portions of the sigmoidal curve. The increase in region II, however, may be small. In region III, where fracture toughness K_c or K_{Ic} controls, substantial differences in crack growth rates occur for different R ratios. The upper transition regions on the curves are shifted to lower ΔK_I values as R, and hence K_{max}, increases. The most commonly used equation depicting mean stress effects in regions II and III is the Forman equation [32]:

$$\frac{da}{dN} = \frac{A(\Delta K)^n}{(1 - R)K_c - \Delta K} \tag{5.21}$$

where A and n are empirical fatigue material constants and K_c is the applicable fracture toughness for the material and thickness. The Forman equation is a modification of the Paris equation.

The effect of mean stress on ΔK_{th} can be substantial, as is indicated in Fig. 5.24 and Table A.4. ΔK_{th} for nine materials with different positive R ratios are included in Table A.4. For R increasing from zero to about 0.8, the threshold ΔK_{th} decreases by a factor of about 1.5 to 2.5. This has the effect of shifting the curve in Fig. 5.23 toward the abscissa by these same factors, which reduces ΔS_{th} for a given crack length by the same factors.

The effect of negative R ratios, which includes compression in the cycle, has not been sufficiently investigated, particularly at the threshold levels. Stephens et al. [33] obtained da/dN versus ΔK for regions II and III for Man-Ten steel as shown in Fig. 5.25, with R ranging from $+0.5$ to -2.0. The scatter band is small except for $R = -2$ at lower crack growth rates. Here, the $R = -2$ curve reversed, because of high crack growth rates immediately following precracking with $R = 0$. The results of many negative R ratio tests on wrought and cast steels, cast irons, and aluminum alloys subjected to constant amplitude conditions in regions II and III indicate

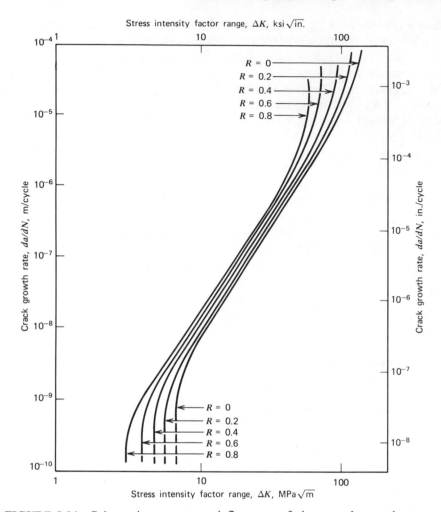

FIGURE 5.24   Schematic mean stress influence on fatigue crack growth rates.

crack growth rates based on ΔK values (which neglect compressive nominal stresses) are similar to $R = 0$ results, or are increased by not more than a factor of 2 [33–35].

   Nelson [36] analyzed region II and III data obtained from Hudson [37] by plotting values of constant da/dN for different R ratios on axes of stress intensity amplitude, $\Delta K/2$, versus the mean stress intensity, K_m. These axes are analogous to the stress axes of S_a versus S_m in the constant life diagrams of Figs. 5.12 to 5.14. Nelson considered both positive and negative K values in calculating $\Delta K/2$. Figure 5.26 shows the results for 7075-T6 aluminum. It should be noted that the trend for a given da/dN curve in going from $R = 0$ to $R = 0.8$ can be estimated by drawing a line from $R = 0$ data to K_c. The 45 degree lines from $R = 0$ to $R = -1$ indicate that compression loadings should have little effect on the constant amplitude fatigue crack growth rates.

5.5.4 Crack Growth Life Integration Example

Let us assume a very wide SAE 1020 cold-rolled plate is subjected to constant amplitude uniaxial cyclic loads that produce nominal stresses varying from S_{max} = 200 MPa (29 ksi) to S_{min} = −50 MPa (−7.3 ksi). The monotonic properties for this steel are S_y = 630 MPa (91 ksi), S_u = 670 MPa (97 ksi). E = 207 GPa (30 × 10⁶ psi), and K_c = 104 MPa$\sqrt{m}$ (95

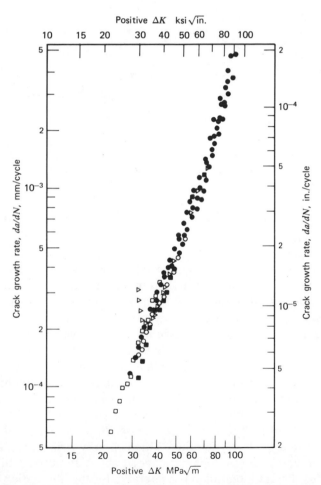

FIGURE 5.25 Positive stress intensity factor range versus crack growth rate with negative R ratios, Man-Ten steel [33] (reprinted by permission of the American Society for Testing and Materials). (□) R = +0.5, (●) R = 0, (○) R = −0.5, (■) R = −1, (△) R = −2.

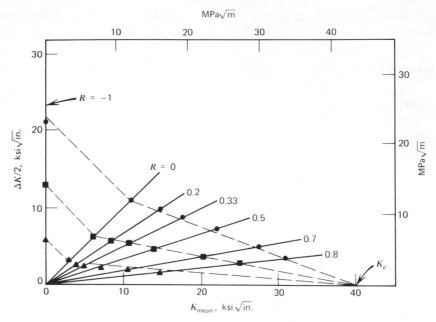

FIGURE 5.26 Mean stress effects on fatigue crack growth rates for 7075-T6 aluminum [36] (reprinted with permission of the Society for Experimental Stress Analysis). Values of da/dN, in./cycle (cm/cycle: (●) 10^{-3} (2.5×10^{-3}), (■) 10^{-4} (2.5×10^{-4}), (▲) 10^{-5} (2.5×10^{-5}).

ksi$\sqrt{\text{in.}}$). What fatigue life would be attained if an initial through the thickness edge crack were no greater than 0.5 mm (0.02 in.) in length?

Before we can solve this problem, several questions must be answered. Namely, what is the applicable stress intensity factor expression for this component and loading? What crack growth rate equation should be used? How do we integrate this equation? What value of ΔK will cause fracture? Does corrosion or temperature play an important part?

In solving this problem let us assume a corrosive environment is not involved and that room temperature prevails. The Paris crack growth rate equation (Eq. 5.17) is often a reasonable expression for region II, and even region III, crack growth behavior. Integration of the Paris equation involves numerical methods unless α from Eq. 5.16 is independent of crack length a. Since K_1 for an infinite plate with a single edge crack has α equal to a constant of 1.12 from Table 4.1b or Fig. 4.5b, it would be desirable to assume the infinite plate K_1 solution. This is a very reasonable assumption as long as the crack length does not exceed about 10 percent of the width (see Fig. 4.5b). Direct integration is preferable for the illustration and hence the infinite plate is assumed. K_{max} initial with $S = 200$ MPa and $a_i = 0.5$ mm is

$$1.12 \, S\sqrt{\pi a} = (1.12)(200 \text{ MPa})\sqrt{\pi (0.0005)} = 9 \text{ MPa}\sqrt{\text{m}}$$

which is above threshold levels and hence the Paris equation is applicable. The final crack length a_f (or a_c is also commonly used) can be obtained from setting K_{max} at fracture equal to K_c.

Thus the following equations apply:

$$\Delta K = S\sqrt{\pi a}\,\alpha \tag{5.22a}$$

$$\frac{da}{dN} = A(\Delta K)^n = A(\Delta S\sqrt{\pi a}\,\alpha)^n$$

$$= A(\Delta S)^n(\pi a)^{n/2}\,\alpha^n \tag{5.22b}$$

$$a_f = \frac{1}{\pi}\left(\frac{K_c}{S_{max}\,\alpha}\right)^2 \tag{5.22c}$$

Integrating Eq. 5.22b,

$$N_f = \int_0^{N_f} dN = \int_{a_i}^{a_f} \frac{da}{A(\Delta S)^n(\pi a)^{n/2}\,\alpha^n}$$

$$= \frac{1}{A(\Delta S)^n(\pi)^{n/2}\,\alpha^n}\int_{a_i}^{a_f}\frac{da}{a^{n/2}} \tag{5.22d}$$

If $n \neq 2$

$$\int_{a_i}^{a_f}\frac{da}{a^{n/2}} = \frac{a^{-(n/2)+1}}{-n/2+1}\Bigg]_{a_i}^{a_f} = \frac{a_f^{(-n/2)+1} - a_i^{(-n/2)+1}}{-n/2+1} \tag{5.22e}$$

Therefore,

$$N_f = \frac{a_f^{(-n/2)+1} - a_i^{(-n/2)+1}}{(-n/2+1)A(\Delta S)^n(\pi)^{n/2}\,\alpha^n} \tag{5.22f}$$

Equation 5.22f is the general integration of the Paris equation when α is independent of crack length a and when n is not equal to 2. This equation is not correct if α is a function of a, which is the usual case.

Since specific crack growth rate data were not given for the SAE 1020 cold-rolled steel, a reasonable first approximation could use the conservative empirical equation 5.18 for ferritic–pearlitic steels suggested by Barsom [31]. Although this equation was developed for $R = 0$, the small compressive stress, 50 MPa, will not have much effect on crack growth and can be neglected and thus $\Delta S = 200 - 0 = 200$ MPa is very reasonable. From Eq. 5.22c

$$a_f = \frac{1}{\pi}\left(\frac{K_c}{S_{max}\,\alpha}\right)^2 = \frac{1}{\pi}\left(\frac{104}{200\times 1.12}\right)^2$$

$$= 0.068\text{ m} = 68\text{ mm} = 2.7\text{ in.} \tag{5.22g}$$

Substitution of the appropriate values into Eq. 5.22f results in

$$N_f = \frac{(0.068)^{-(3/2)+1} - (0.0005)^{-(3/2)+1}}{(-3/2 + 1)(6.9 \times 10^{-12})(200)^3(\pi)^{3/2}(1.12)^3}$$

$$= \frac{(0.068)^{-0.5} - (0.0005)^{-0.5}}{-2.16 \times 10^{-4}} = 189,000 \text{ cycles}$$

Now let us assume the fracture toughness, K_c, was incorrect by a factor of ±2, that is, K_c = 208 MPa$\sqrt{m}$ (180 ksi$\sqrt{in}$.) or 52 MPa$\sqrt{m}$ (42.5 ksi$\sqrt{in}$.). The final fracture length from Eq. 5.22c would result in a_f = 270 mm (10.8 in.) and 17 mm (0.68 in.), respectively. The final life N_f from Eq. 5.22f would be 198,000 and 171,000 cycles, respectively. Thus increasing or decreasing the fracture toughness by a factor of 2 caused increases or decreases in the final crack length by a factor of 4, respectively. However, changes in fatigue crack growth life were less than 10 percent, which are very small differences. If the initial crack length a_i were 2.5 mm (0.1 in.), the life would have been only 75,000 cycles instead of 189,000 cycles for the original problem with a_i equal to 0.5 mm (0.02 in.). This illustrates the importance of minimizing initial flaw or crack lengths to obtain long fatigue life and that appreciable changes in fracture toughness will alter final crack lengths but may not have appreciable effects on fatigue life. High fracture toughness in fatigue design, however, is still very desirable because of the randomness of many load histories and the larger crack lengths before fracture, which permits much better inspection success.

5.6 SCATTER AND STATISTICAL ASPECTS IN FATIGUE

Scatter in fatigue testing or in component fatigue life is a very important consideration in using test data. Sinclair and Dolan [38] performed an extensive statistical fatigue study involving 174 so-called identical highly polished unnotched 7075-T6 aluminum alloy specimens. They used six different alternating stress levels under fully reversed (S_m = 0) conditions. Figure 5.27 shows histograms of fatigue life distribution of 57 specimens tested at a 207 MPa (30 ksi) stress level. A stepped histogram can be replaced by a smooth curve (dashed lines in Fig. 5.27) known as a frequency distribution curve. When normalized to unit area, this becomes the probability density function. The curve on the left, which is based on cycles to failure, is skewed and hence is not a normal or Gaussian distribution. The curve on the right is based on the logarithm of cycles to failure and reasonably approximates a normal or Gaussian distribution. Figure 5.28 includes the entire 174 tests plotted on log-normal probability paper for each value of S_a. If the data are truly log-normal, the data points for each value of S_a will be on a straight line. As seen, a log-normal distribution for the actual data region at each stress level appears reasonable. Based on these and many other statistical test results, a log-normal distribution of fatigue life is often assumed in fatigue design.

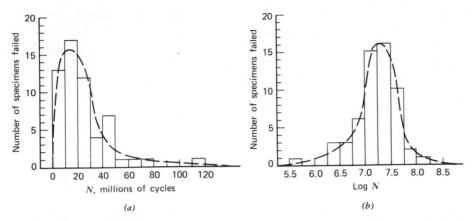

FIGURE 5.27 Histograms showing fatigue life distributions for 57 specimens of 7075-T6 aluminum alloy tested at 207 MPa (30 ksi) [38] (reprinted with permission of the American Society of Mechanical Engineers).

An analysis of Fig. 5.28 reveals that less scatter occurred at the higher stress levels, as indicated by the steeper slopes. At the highest stress level the life varied from about 1.5×10^4 to 2×10^4 cycles, or a factor of less than 2. At the lowest stress level the life varied from about 2×10^6 to 7×10^7 cycles, or a factor of about 35. Factors of 100 in life are not uncommon for very low stress level fatigue tests. Scatter is usually greater in unnotched polished specimens than in notched or cracked specimens. Scatter can be attributed to testing techniques, specimen preparation, variations in the

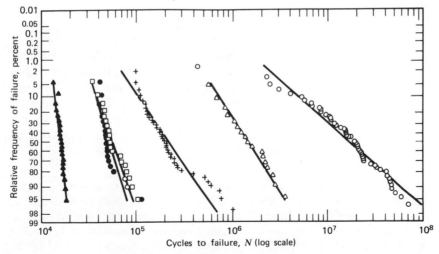

FIGURE 5.28 Log-normal probability plot at different stress levels for 7075-T6 aluminum alloy [38] (reprinted with permission of the American Society of Mechanical Engineers). (▲) 430 MPa (62.5 ksi), (●) 345 MPa (50 ksi), (□) 310 MPa (45 ksi), (+) 275 MPa (40 ksi), (△) 240 MPa (35 ksi), (○) 207 MPa (30 ksi).

material, and variability of the fatigue mechanisms. The greater scatter at low stress levels in these smooth unnotched specimens can be attributed to the greater percentage of life needed to initiate small microcracks and then macrocracks. At higher stress levels a greater percentage of the fatigue life involves propagation of macrocracks. Tests involving only fatigue crack growth under constant amplitude conditions usually show scatter factors of 2 or 3 or less for identical tests. Thus the greatest scatter in fatigue involves the initiation of microcracks and small macrocracks. In notched specimens and components, cracks form quicker, and subsequently a greater proportion of the total fatigue life involves crack propagation that has less scatter.

Weibull* distributions are often used in preference to the log-normal distribution to analyze probability aspects of fatigue results. Weibull developed this engineering approach while determining fatigue lives of ball bearings [39, 40]. Both two and three parameter Weibull distribution functions exist, but the two parameter function is most frequently used in fatigue design and testing. It assumes the minimum life N_0 of a population is zero, while the three parameter function defines a finite minimum life other than zero. The three parameter Weibull model is

$$F(N) = 1 - \exp \{-[(N - N_0)/(\theta - N_0)]^b\} \tag{5.23}$$

where $F(N)$ = fraction failed in time or cycles N
$\quad\quad N_0$ = minimum time or cycles to failure
$\quad\quad \theta$ = characteristic life (time or cycles when 63.2 percent have failed)
$\quad\quad b$ = Weibull slope or shape parameter

The terms N_0, θ, and b are the three Weibull parameters. The two parameter Weibull model has $N_0 = 0$ and hence

$$F(N) = 1 - e^{-(N/\theta)^b} \tag{5.24}$$

Weibull probability paper is available on which percent failed is plotted against time or cycles to failure. Figure 5.29 is a three parameter Weibull distribution plot of 1814 fatigue tests of mild steel, thin sheet specimens tested at one constant amplitude condition [41]. Data points represent only partial data from the extensive test program. Several points have been labeled for better clarity, indicating the number of specimens failed at that life. The minimum life, N_0, of 80,987 cycles must be added to the lives in

---

* Waloddi Weibull (1887). He received his B.A. in 1912, M.S. in 1924, and Ph.D. in 1932 from the University of Uppsala, Sweden. He was Director of research at NKA Ball Bearing Co., Goteborg, Sweden from 1916 to 1922. Weibull was then employed as Professor of Mechanical Engineering, Royal Institute of Technology, Stockholm from 1923 to 1941 and Professor of Applied Physics from 1941 to 1953. He is a member of The Royal Swedish Academies of Science, Engineering Science, Military Science, and Naval Science and has remained active in engineering and reliability ever since retiring from his university position in 1953.

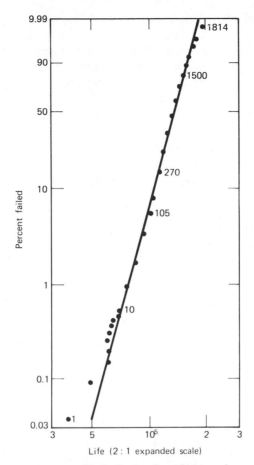

FIGURE 5.29   Three parameter Weibull plot for 1814 specimens of mild steel [41].

Fig. 5.29 to obtain total life. Additional numerical data on scatter in fatigue and its treatment by statistical methods is given in Appendix B. Reference 42 also includes information on statistical analysis of fatigue data.

On Weibull paper most data can be plotted as straight lines by judicious choice of N_0. However, as with all distribution function models, the tails at each end may deviate from the model. This is seen in Fig. 5.29. The slope, b, of the straight line gives a measure of the scatter and the shape of the distribution. For $b = 3.5$, the Weibull distribution function is approximately normal or Gaussian. The coefficient of variation (standard deviation/mean) is approximately $C = 1/b$ for the two parameter Weibull distribution.

Data in Fig. 5.28 for 7075-T6 aluminum are quite comprehensive, yet do not provide adequate engineering information for fatigue life decisions based on real-life desired probability of failure. Data do not extend past 2 percent probability of failure. Engineers are most interested in 0.01 percent probability of failure or less. This means extrapolation is needed, for which no accurate method nor mathematical justification exists. Only engineering judgment justifies extrapolation. For instance, in Fig. 5.28, if the six stress level curves were extrapolated to desirable low probabilities of failure, they would intersect, which implies a higher stress level would give longer fatigue life than a lower stress level, which is unreasonable. Great caution must be used in extrapolating fatigue data to low probabilities of failure.

To be able to estimate low probabilities of failure Abelkis [43] has pooled data from more than 6600 specimens of aluminum alloys tested in 1180 groups or samples. He derived a three term exponential equation:

$$F(z) = A_1 e^{d_1 z} + A_2 e^{d_2 z} + A_3 e^{d_3 z} \quad \text{for} \quad z < 0 \tag{5.25}$$

where $z = [\log N - \log (\text{median } N)]/D$
N = number of cycles to failure
D = standard deviation of log N
F = probability of failure ($F = 50$ percent for $z = 0$)

The coefficients and exponents are

$$A_1 = 1.687 \sqrt{D} \qquad\qquad d_1 = 1.3 + 0.86\sqrt{D}$$
$$A_2 = 0.115 \qquad\qquad\quad d_2 = 0.28 + 0.44\sqrt{D}$$
$$A_3 = 0.485 - 1.687\sqrt{D} \qquad d_3 = 1.09 + 2.16\sqrt{D}$$

Near the mean this distribution is almost the same as the log-normal distribution, but for low probabilities of failure it predicts a much lower life than the log-normal distribution. For example, Abelkis shows test data from 2103 specimens in 375 groups. The standard deviation of log N is $D = 0.175$. Both the log-normal distribution and his distribution show 16 percent failed at $z = -1$ or at a life $1/10^{0.175} = 67$ percent of the median life. Table 5.1 shows values that apply for lower fractions of failures. Use of the log-normal distribution would have predicted five times longer life than Abelkis for the first five failures of 10,000 specimens. The formula and data from Abelkis [43] are quoted not to recommend them as the best distribution

TABLE 5.1   Expected Fractions of Mean Life for Various Probabilities of Failure

Fraction Failed, percent	Expected z and Percent of Median Life			
	From Log Normal		From Albelkis	
	z	Fraction of Med. Life, percent	z	Fraction of Med. Life, percent
2	-2	45	-2.2	41
0.2	-2.9	31	-4.7	15
0.05	-3.3	26	-7.3	5

function, but as an example of a function that is more realistic than the log-normal function.

5.7 DOS AND DON'TS IN DESIGN

1. Do consider that the fully reversed fatigue limit, S_f, for components can vary from about 1 to 70 percent of the ultimate tensile strength and that the engineer can substantially influence this value by proper design decisions.

2. Don't use the median constant amplitude fatigue material properties in Tables A.1 and A.2 for design unless they are modified for lower probabilities of failure and various conditions, such as corrosive and temperature environments, surface and size effects, self-stresses, and notches. This applies to most other published fatigue material properties too.

3. Don't neglect the advantages of compressive mean or self-stresses (residual stresses) in improving fatigue life.

4. Do recognize that strain–life fatigue data of smooth uniaxial specimens are based on cycles to "failure" where failure represents formation of cracks between 0.25 and 5 mm (0.01 and 0.2 in.) in depth, which may or may not have caused fracture. The crack depths at fracture are controlled by S_{max} and fracture toughness, K_c.

5. Don't neglect statistical considerations in fatigue design. The log-normal and two or three parameter Weibull life distribution functions can be very helpful, but extrapolation of any distribution function to extreme probabilities has no mathematical nor experimental justification except engineering judgment.

6. Do note that most fatigue crack growth occurs in mode I even under mixed-mode conditions, and hence the opening mode stress intensity factor range ΔK_I is the predominant controlling factor in fatigue crack growth.

7. Do use LEFM principles in fatigue crack growth life predictions even in low strength materials; crack tip plasticity can be small even in low strength materials under fatigue conditions.

8. Do consider the possibility of inspection before fracture. High fracture toughness materials may not provide appreciable increases in fatigue crack growth life, but they do permit longer cracks before fracture, which makes inspection more reliable.

REFERENCES FOR CHAPTER 5

1. L. Spangenburg, *The Fatigue of Metals Under Repeated Strains*, D. Van Nostrand Co., Princeton, NJ, 1876.

2. P. W. Bridgman, "The Stress Distribution at the Neck of a Tension Specimen," *Trans. ASM,* Vol. 32, 1944, p. 553.

3. J. B. DeJong, D. Schütz, H. Lowak, and J. Schütz, "A Standardized Load Sequence for Flight Simulation Tests on Transport Aircraft Wing Structures," NLR, Report No. TR73029U, The Netherlands.

4. R. M. Wetzel, Ed., *Fatigue Under Complex Loading: Analysis and Experiments,* SAE, 1977.

5. J. M. Barsom, "Fatigue–Crack Growth Under Variable–Amplitude Loading in ASTM A514-B Steel," *Progress in Flaw Growth and Fracture Toughness Testing,* ASTM STP 536, 1973, p. 147.

6. "Standard Recommended Practice of Constant Amplitude Axial Fatigue Tests of Metallic Materials," ASTM Designation E466-76.

7. "Standard Recommended Practice for Verification of Constant Amplitude Dynamic Loads in an Axial Load Fatigue Testing Machine," ASTM Designation E467-76.

8. S. R. Swanson, Ed., *Handbook of Fatigue Testing,* ASTM STP 566, 1974.

9. "Tentative Test Method for Constant-Load-Amplitude Fatigue Crack Growth Rates Above 10^{-8} m/cycle," ASTM Designation E647-78T.

10. "Standard Definitions of Terms Relating to Fatigue Testing and Statistical Analysis of Fatigue Data," ASTM Designation E206-72.

11. *Aluminum Standards and Data 1976,* The Aluminum Association, New York, 1976.

12. R. C. Juvinal, *Engineering Considerations of Stress, Strain and Strength,* McGraw-Hill Book Co., New York, 1967.

13. J. E. Shigley, *Mechanical Engineering Design,* 2nd ed., McGraw-Hill Book Co., New York, 1972.

14. P. G. Forrest, *Fatigue of Metals,* Pergamon Press, London, 1962.

15. D. K. Bullens, *Steel and Its Heat Treatment,* John Wiley and Sons, New York, Vol. 1, 1938, p. 37.

16. R. W. Landgraf, "The Resistance of Metals to Cyclic Deformation," *Achievement of High Fatigue Resistance in Metals and Alloys,* ASTM STP 467, 1970, p. 3.

17. H. J. Grover, *Fatigue of Aircraft Structures,* NAVAIR 01-1A-13, 1966.

18. *Mil-Handbook 5,* Department of Defense, Washington, D.C.

19. G. Sines, "Failure of Materials Under Combined Repeated Stresses with Superimposed Static Stresses," NACA TN 3495, 1955.

20. H. O. Fuchs, "A Set of Fatigue Failure Criteria," *Trans. ASME, J. Basic Eng.,* Vol. 87, June 1965, p. 333.

21. "Technical Report on Fatigue Properties," SAE J1099, 1975.

22. O. H. Basquin, "The Exponential Law of Endurance Tests," *Proc. ASTM,* Vol. 10, Part II, 1910, p. 625.

23. J. F. Tavernelli and L. F. Coffin, Jr., "Experimental Support for Generalized Equation Predicting Low Cycle Fatigue," *Trans. ASME, J. Basic Eng.,* Vol. 84, No. 4, Dec. 1962, p. 533.

24. S. S. Manson, discussion of reference 23, *Trans. ASME J. Basic Eng.*, Vol. 84, No. 4, Dec. 1962, p. 537.

25. J. A. Graham, Ed., *Fatigue Design Handbook*, SAE, 1968.

26. S. S. Manson, "Fatigue: A Complex Subject—Some Simple Approximations," *Exp. Mech.*, Vol. 5, No. 7, July 1965, p. 193.

27. K. N. Smith, P. Watson, and T. H. Topper, "A Stress–Strain Function for the Fatigue of Metals," *J. Mater.*, Vol. 5, No. 4, Dec. 1970, p. 767.

28. P. C. Paris and F. Erdogan, "A Critical Analysis of Crack Propagation Laws," *Trans. ASME, J. Basic Eng.*, Vol. 85, No. 4, 1963, p. 528.

29. D. W. Hoeppner and W. E. Krupp, "Prediction of Component Life by Application of Fatigue Crack Growth Knowledge," *Eng. Fract. Mech.*, Vol. 6, 1974, p. 47.

30. *Damage Tolerant Design Handbook*, Metals and Ceramics Information Center, Battelle Labs, Columbus, OH, 1975.

31. J. M. Barsom, "Fatigue-Crack Propagation in Steels of Various Yield Strengths," *Trans. ASME, J. Eng. Ind.*, Ser. B, No. 4, Nov. 1971, p. 1190.

32. R. G. Forman, V. E. Kearney, and R. M. Engle, "Numerical Analysis of Crack Propagation in Cyclic-Loaded Structures," *Trans. ASME, J. Basic Eng.*, Vol. 89, No. 3, 1967, p. 459.

33. R. I. Stephens, E. C. Sheets, and G. O. Njus, "Fatigue Crack Growth and Life Predictions in Man–Ten Steel Subjected to Single and Intermittent Tensile Overloads," *Cyclic Stress–Strain and Plastic Deformation Aspects of Fatigue Crack Growth*, ASTM STP 637, 1977, p. 176.

34. R. I. Stephens, "Fatigue Crack Growth Retardation: Fact and Fiction," *Proceed. 9th ICAF Symp., Darmstadt, Ger.*, 1977, p. 5.1/1.

35. R. I. Stephens, P. H. Benner, G. Mauritzson, and G. W. Tindall, "Constant and Variable Amplitude Fatigue Behavior of Eight Steels," *J. Test. Eval.*, Vol. 7, No. 2, March 1979, p. 68.

36. D. V. Nelson, "Review of Fatigue-Crack-Growth Prediction Methods," *Exp. Mech.*, Vol. 17, No. 2, Feb. 1977, p. 41.

37. C. M. Hudson, "The Effect of Stress Ratio on Fatigue Crack Growth in 7075-T6 and 2024-T3 Aluminum Alloy Specimens," NASA TN-D-5390, August 1969.

38. G. M. Sinclair and T. J. Dolan, "Effect of Stress Amplitude on Statistical Variability in Fatigue Life of 75S-T6 Aluminum Alloy," *Trans. ASME*, Vol. 75, 1953, p. 867.

39. W. Weibull, "A Statistical Distribution Function of Wide Applicability," *J. Appl. Mech.*, Sept. 1951, p. 293.

40. W. Weibull, *Fatigue Testing and Analysis of Results*, Pergamon Press, London, 1961.

41. W. E. Hering and C. W. Gadd, "Experimental Study of Fatigue Life Variability," General Motors Corp., Research Publication, GMR-555, 1966.

42. *A Tentative Guide for Fatigue Testing and the Statistical Analysis of Fatigue Data*, ASTM STP 91A, 1958.

43. P. R. Abelkis, "Fatigue Strength Design and Analysis of Aircraft Structure, Part I: Scatter Factors and Design Charts," AFFDL-TR 66-197, June 1967.

PROBLEMS FOR CHAPTER 5

1. What difference exists between true stress, σ, and engineering stress, S, and between true strain, ϵ, and engineering strain, e, if engineering strain is 2 percent? Can this difference be neglected for engineering design purposes?

2. If a structural component is subjected to repeated stress cycles where $S_{max} = 400$ MPa, $S_{min} = -600$ MPa, determine the following; (a) S_m, (b) S_a, (c) ΔS, (d) R, and (e) A.

3. Fatigue testing can take an appreciable amount of time. Calculate the number of days it would take to apply 10^6, 10^7, and 10^8 cycles for test frequencies of:

    (a) 1 Hz (approximate speed of Wöhler's original work.)*
    (b) 30 Hz (speed of many common test machines).
    (c) 150 Hz (speed of some rotating beam test machines).

    Estimate the number of cycles the following items must endure during their expected life time: bicycle pedal shaft, truck engine valve spring, home light switch, automobile brake pedal.

4. Show that the stress in a rotating beam fatigue specimen is sinusoidal as a function of *time* when rotated at a constant angular velocity.

5. For the following $R = -1$ AISI 1090 steel test data, plot two S–N curves, one using rectangular coordinates and one using semilog (≥ 3 cycle) graph paper. Draw median and lower bound S–N curves. Can you estimate the S–N curve for 0.01 percent failures? What advantages or disadvantages exist in the two coordinate systems? From the S–log N curve determine: (a) fatigue limit, (b) fatigue strength at 5×10^5 cycles, and (c) fatigue life at $S_a = 260$ MPa. Comment on the scatter and how you handled it for parts a through c.

S_a, MPa	Cycles to Failure	S_a, MPa	Cycles to Failure
340	15×10^3	250	301×10^3
300	24×10^3	235	290×10^3
290	36×10^3	230	361×10^3
275	80×10^3	220	881×10^3
260	177×10^3	215	1.3×10^6
255	162×10^3	210	2.6×10^6

    The following stress levels had $> 10^7$ cycles without failure: 210, 210, 205, 205, and 205.

6. Why should unnotched axial fatigue limits be 10 to 25 percent less than unnotched fatigue limits obtained from rotating bending? List several contributing factors.

7. A fatigue life test program on a new component consisted of 10 com-

* One of Wöhler's tests ran 1.3×10^8 cycles.

ponents subjected to the same load spectrum. The components failed in the following numbers of hours in ascending order: 75, 100, 130, 150, 185, 200, 210, 240, 265, and 300. Obtain the percent median rank values for the 10 failures. Plot the percent median rank values versus hours to failure on Weibull paper. Draw the best fit (usually a straight line) curve through the data and find:

(a) 10 percent (B_{10}) expected failure life.
(b) 50 percent (B_{50}) expected failure life.
(c) What percentage of the population will have failed in 230 hours.
(d) Consider what the 1.0 or 0.1 percent expected life would be.

8. A smooth uniaxial rod with a cross-sectional area of 0.003 m² is made from a material with S_f = 300 MPa under $R = -1$ conditions. S_u = 700 MPa and RA = 20 percent. If the rod is subjected to a mean force of 180 kN, what is the allowable alternating force that will not cause failure in 10^6 cycles according to the Gerber criteria?

9. A smooth uniaxial bar is subjected to a minimum stress of 35 MPa in compression. S_f for $R = -1$ was 220 MPa and S_u was 500 MPa. Using the modified Goodman criteria, determine the maximum tensile stress the bar will withstand without failure in 10^6 cycles.

10. Using the data in Table A.2 and E for steel equal to 200,000 MPa, determine the total, elastic, and plastic strain amplitudes for smooth uniaxial test specimens of Man-Ten and RQC-100 steels for life to (a) 10^3 cycles and (b) 10^5 cycles (note one cycle equals two reversals).

11. Repeat problem 10 using the Manson method of universal slopes and comment on any differences with problem 10 and their significance.

12. If $\Delta\epsilon/2 = 0.01$, find the number of cycles to failure for Man-Ten and RQC-100 steels using Eq. 5.10. How do these results compare with Fig. 5.18?

13. Examine Table A.2 to see how ϵ_f' and ϵ_f compare and how σ_f' and σ_f compare.

14. If $\sigma_m = +\sigma_f'/4$ what approximate effect does this have on Man-Ten steel strain amplitude for 10^3 and 10^5 cycles. Repeat for $\sigma_m = -\sigma_f'/4$. Compare results with problem 10. Where does σ_m have its greatest influence?

15. The following crack growth data were obtained with SAE 0030 cast steel at $-34°C$ using compact specimens with $H/W = 0.49$, $W = 64.8$ mm, thickness $B = 8.2$ mm, $P_{max} = 6.14$ kN, $P_{min} = 0.089$ kN. K_I for this specimen is given by

$$K_I = \frac{P\sqrt{a}}{Bw}\left[30.96 - 195.8\left(\frac{a}{w}\right) + 730.6\left(\frac{a}{w}\right)^2 \right.$$
$$\left. - 1186.3\left(\frac{a}{w}\right)^3 + 754.6\left(\frac{a}{w}\right)^4\right]$$

Plot a curve of da/dN versus ΔK_{I}. Reduce the data using a numerical scheme. See ASTM tentative test method E647-78T. Obtain the Paris equation coefficient A and exponent n.

Cycles	Crack Length, (mm)	Cycles	Crack Length, (mm)
0	22.0	751,200	27.7
37,100	22.1	801,600	28.9
68,300	22.3	863,400	29.8
111,200	22.5	908,800	30.9
173,500	22.7	942,900	32.0
238,000	23.0	978,000	33.5
286,400	23.3	1,006,700	35.1
371,000	23.75	1,028,900	36.55
446,200	24.1	1,059,700	39.2
501,200	24.9	1,069,400	42.0
560,900	25.8	1,073,200	44.1
691,800	27.0	1,074,900	45.0

16. A uniaxial loaded very wide sheet of low carbon steel, $S_y = 350$ MPa, $S_u = 550$ MPa, $K_c = 110$ MPa$\sqrt{m}$, $\Delta K_{\text{th}}]_{R=0} = 5$ MPa$\sqrt{m}$ is subjected to the following load spectrum.

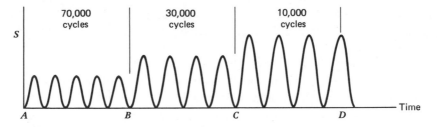

The three stress levels $S_{\max}$ are 150, 200, and 250 MPa, $R = 0$. The initial center crack length $2a = 5.0$ mm. Assume Barsom's crack growth expressions are applicable. Determine:

    (a) K_{I} at time A and compare with ΔK_{th} and K_c.
    (b) What value of $S_{\max}$ will not cause crack growth at time A?
    (c) The crack length at time B, C, and D.
    (d) K_{I} at B, C, and D and compare with K_c.

17. Conversion of the exponential Paris equation for fatigue crack growth rules from SI units to American/British units or vice versa has too frequently been done incorrectly. Verify that the conversions of Eqs. 5.18 to 5.20 have been done correctly.

CHAPTER 6

NOTCHES AND
THEIR EFFECTS

Notch effects have been a key problem in the study of fatigue since more than 100 years ago, when Wöhler showed that adding material to a railway axle might make it weaker in fatigue. He stated that the radius at the shoulder between a smaller and a larger diameter is of prime importance to the fatigue life of axles and that fatigue cracks will start at the transition from a smaller to a larger section.

Notches cannot be avoided in structures and machines. A bolt has notches in the thread roots and at the transition between head and shank. Rivet holes in sheets, welds on plates, keyways on shafts, all are notches. Although notches can be very dangerous they can often be rendered harmless by suitable treatment. For example, Harris [1] showed that a notch that reduced fatigue strength to 60 percent of the unnotched value became perfectly harmless when it was shot-peened (Fig. 6.1).

To understand the effects of notches and the means to overcome them one must consider five parameters in addition to the behavior of smooth specimens of the same material:

Concentrations of stress and of strain.
Stress gradients.
Mean stress effects and self stresses.
Local yielding.
Development and growth of cracks.

In some cases one of the five parameters by itself may explain the difference in behavior between a smooth part and a notched part that has equal cross section at the root of the notch. Even when several parameters are involved it is often possible to invent "variable constants" or "notch factors" that correlate the test results. We intend to avoid the use of adjustable factors and must therefore consider the effects of all five parameters. In the end this will be less difficult than the adjustment of a single notch factor. In this chapter we discuss concentrations of stresses and the development of small

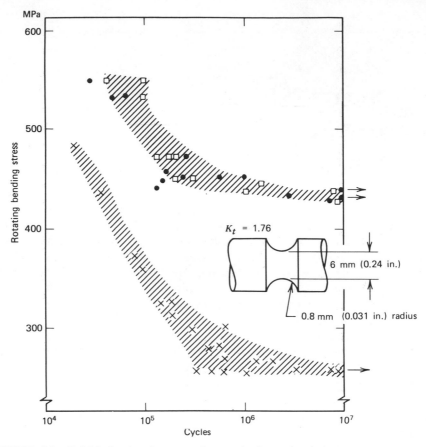

FIGURE 6.1 *S–N* behavior for smooth, notched unpeened, and notched peened specimens [1] (reprinted with permission of Pergamon Press). (●) Smooth, polished, (×) notched, (□) notched, shot-peened.

cracks. Local yielding and self-stresses are discussed in Chapter 7. The growth of cracks has been discussed in Chapter 5. The sum of all these effects is the subject of Chapter 8.

6.1   CONCENTRATIONS AND GRADIENTS OF STRESS AND STRAIN

Notches concentrate stresses and strains. The degree of concentration is a factor in the fatigue strength of notched parts. It is measured by the elastic stress concentration factor K_t. Figure 6.2 shows a sheet of rubber with a round hole, with and without the loading that produces strain. It can be seen that at the edge of the sheet the strain is quite uniform. But on the sides of the hole the strain is larger than in the rest of the sheet. The elastic strain concentration factor or stress concentration factor K_t is defined as the ratio of maximum strain (at the side of the hole) to "nominal strain" (force divided by area divided by Young's modulus).

$$K_t = \frac{\sigma}{S} = \frac{\epsilon}{e} \qquad \text{as long as} \qquad \frac{\sigma}{\epsilon} = \text{const} = E \qquad (6.1)$$

K_t depends on the ratio of hole diameter to sheet width. Figure 6.3 shows K_t plotted versus the ratio of hole diameter to sheet width. Two curves are shown. In the upper curve the nominal stress is defined as load divided by total area (width × thickness). In the lower curve the nominal stress is defined as load divided by net area, that is, the area remaining after the hole has been cut out. In this book we use the net area to define the nominal stress when using stress concentration factors unless otherwise stated. However, in calculating the stress intensity factor from the nominal stress we calculate the latter as if the crack did not exist. In calculating joint efficiencies we compare stresses to those in the gross section without holes.

Figure 6.4 shows the stresses near a circular hole in the center of a wide sheet in tension. This problem has been investigated and solved as early as

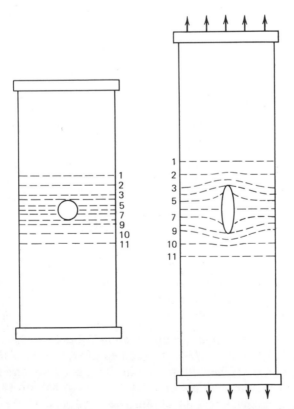

FIGURE 6.2   Concentrations of strain near a hole in a stretched sheet of rubber.

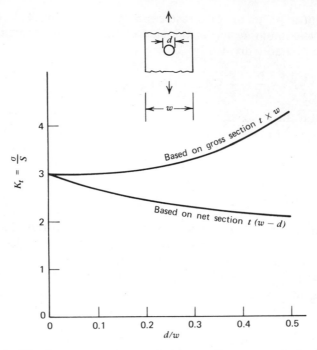

FIGURE 6.3 Elastic stress concentration factor for a central circular hole in a sheet.

1898 by Kirsch [2]. The following equations for the axial stresses σ_y and the transverse stresses σ_x on a transverse line through the center of the hole are quoted from Grover [3].

$$\frac{\sigma_y}{S} = 1 + 0.5 \left(\frac{r}{x}\right)^2 + 1.5 \left(\frac{r}{x}\right)^4 \qquad (6.2)$$

$$\frac{\sigma_x}{S} = 1.5 \left(\frac{r}{x}\right)^2 - 1.5 \left(\frac{r}{x}\right)^4 \qquad (6.3)$$

where S = nominal stress = load/area
σ_y = axial stress
σ_x = transverse stress
x = distance from center of hole
r = radius of hole

Values of σ_y/S from this equation are given in Table 6.1, where d = distance from edge of hole ($d/r = x/r - 1$). Values of σ_y/S and σ_x/S are plotted versus x/r in Fig. 6.4. The bottom row in Table 6.1 gives the relative stress increment, that is, the multiple or fraction of nominal stress that must be added to the nominal stress to obtain the local stress. We see that this increment decreases quite rapidly as the distance from the edge of the hole

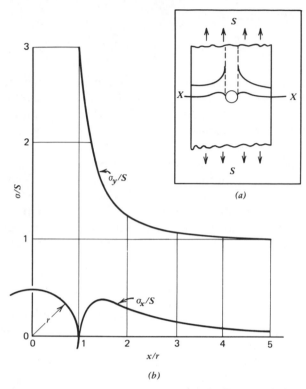

FIGURE 6.4   Stress distribution along the section X–X near a hole in a wide sheet.

is increased. At a distance of $0.25r$ the increment is only about 50 percent of the value at the edge of the hole. At $0.5r$ it is only about 25 percent. At r it is only a little over 10 percent. In other words, the stress at the edge of the hole is three times the nominal stress, but at a distance equal to the diameter it is only about 7 percent higher then the nominal stress.

The slope of the σ_y–x curve at the edge of the hole is another measure of the rapid decrease of stress as we move away from the edge of the hole. We define the relative stress gradient

$$Q = \frac{1}{\sigma_{\max}} \frac{d\sigma}{dx} \qquad (6.4)$$

TABLE 6.1   Stress Distribution Near a Hole in a Sheet in Tension

d/r	0	0.1	0.25	0.5	1	1.6	2	3	4
x/r	1	1.1	1.25	1.5	2	2.6	3	4	5
σ_y/S	3	2.44	1.94	1.52	1.22	1.11	1.07	1.04	1.02
σ_x/S	0	0.22	0.35	0.37	0.28	0.19	0.15	0.09	0.06
$(\sigma_y/S) - 1$	2	1.44	0.94	0.52	0.22	0.11	0.07	0.04	0.02

From Eq. 6.2 this calculates as

$$Q = -\frac{7}{3r} = -\frac{2.33}{r}$$

The rapid decrease of stress with increasing distance from the notch and the existence of biaxial or triaxial states of stress at a small distance from the notch are typical of stress concentrations. They explain why we cannot expect to predict the behavior of notched parts with great accuracy by applying stress concentration factors to the fatigue strength values obtained from smooth parts. Numerical values of stress gradients or simple design formulas for stress distribution near notches are not readily available in the literature. Table 6.2 gives some values of relative stress gradients. A value of $2.5/r$ is a good general approximation. For deep narrow notches with semicircular ends a formula analogous to linear elastic fracture mechanics formulas has been given for the stress distribution [5]. It is

$$\sigma_y = \sigma_{max} \left(\frac{0.5r}{0.5r + d} \right)^{1/2} \tag{6.5}$$

where d is the distance from the edge of the notch of radius r.

The stress concentration produced by a given notch is not a unique number. It depends on the mode of loading and on the type of stress that is used to calculate K_t. For instance, for the circular hole in a wide sheet, we have the following concentration factors:

In tension	3
In biaxial tension	2
In shear	4 based on maximum tension
	2 based on maximum shear

Elastic stress concentration factors are obtained from the theory of elasticity, from finite element computations, or from measurements with strain

TABLE 6.2   Relative Stress Gradients Q Near Notches of Radius r [4]

$$Q = \frac{1}{\sigma_{max}} \frac{d\sigma}{dx}$$

Small central hole in wide sheet	Axial load	$2.33/r$
Fillet in shaft	Axial load	$2/r$
Diameters D and d	Bending	$2/r + 4/(D + d)$
	Torsion	$1/r + 4/(D + d)$
Central hole in shaft	Axial load	$2.5/r$
Diameter of shaft $= D$	Bending	$4/r + 2/D$
	Torsion	$3/r + 2/D$

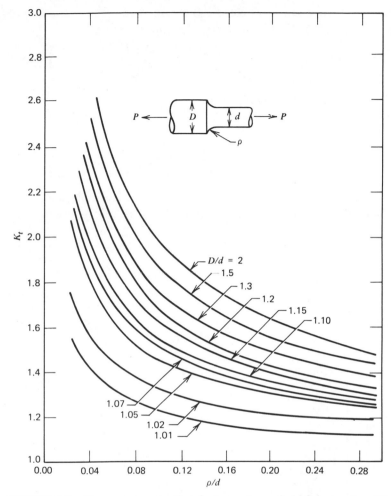

FIGURE 6.5   Stress concentration factors, stepped shaft in tension [3].

gages or photoelastic models. It is usually less important to know their precise magnitude than to know where the most dangerous concentrations exist in a part. Charts of stress concentration factors are available in the literature [6]. Figures 6.5 through 6.7 show examples of such charts for stepped shafts in tension, bending, and torsion [3].

For qualitative estimates some engineers like to use an imperfect analogy between stresses or strains and liquid flow. Restrictions or enlargements in a pipe produce local increases in flow velocity somewhat similar to the local increases in stresses produced by changes in cross section. The designer will try to "streamline" the contours of parts as indicated in Fig. 6.8. Consider for instance an elliptic hole in a wide sheet. Placed lengthwise with the forces or flow it produces less stress concentration and less flow

interference than when it is placed crosswise. The formula for K_t produced
by an elliptic hole with principal axes $2a$ and $2b$ is:

$$K_t = 1 + 2\frac{b}{a} = 1 + 2\sqrt{\frac{b}{r}} \qquad (6.6)$$

where b is the axis transverse to the tension force and r is the radius of
curvature at the endpoint of b. With an ellipse 30 mm long and 10 mm wide
the stress concentration is:

$$K_t = 7 \text{ if placed crosswide}$$
$$K_t = 1.67 \text{ if placed lengthwise}$$

(The radii of curvature are 1.7 and 45 mm). This example points to a way
of mitigating stress concentrations. If we need a hole, we can reduce the

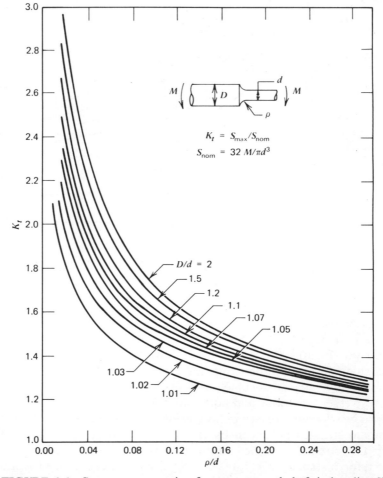

FIGURE 6.6   Stress concentration factors, stepped shaft in bending [3].

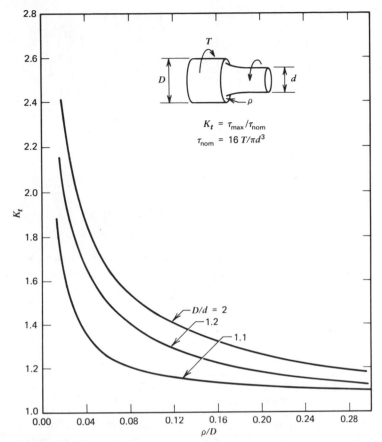

FIGURE 6.7 Stress concentration factors, stepped shaft in torsion [3].

stress concentration by elongating it to an ellipse or by adding smaller relief holes above and below it as shown in Fig. 6.9 [7].*

6.2 THE FATIGUE NOTCH FACTOR

When the theory of elasticity was used to compute stress concentrations there was hope that the fatigue strength of a notched component could be predicted as the strength of a smooth component divided by a factor computed from the theory. The facts, however, are different. For instance, Fig. 6.10 shows the fatigue strengths of smooth and notched specimens of aluminum alloy 7075-T6 at 10^4 and at 10^7 cycles plotted versus the mean stress [3]. The elastic stress concentration factor was 3.4. The ratio of smooth to

---

* Additional examples of mitigating stress concentrations are given in Fig. 6.13

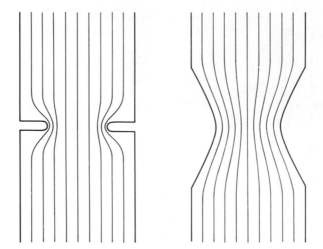

FIGURE 6.8 The crowding of flow lines near obstructions helps to visualize the concentration of strains near notches. The large section and the small section are the same in both cases, but the transitions are different.

notched fatigue strengths, based on the ratio of alternating stresses is called K_f. In Fig. 6.10 it is as shown below.

	At 10^4 cycles	At 10^7 cycles
At zero mean stress	51/22 = 2.3	22/9 = 2.3
At 172 MPa (25 ksi) mean stress	42/13 = 3.2	17/3 = 5.7

Obviously the fatigue notch factor K_f is not necessarily equal to the elastic stress concentration factor. It may be less or it may be more. Within

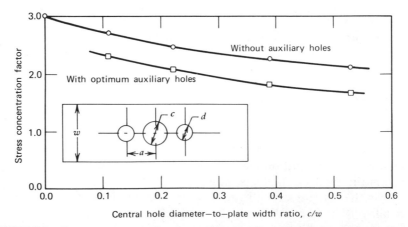

FIGURE 6.9 Stress concentration factor, K_t, as a function of central hole diameter-to-plate width ratio, c/w, for a uniaxially loaded plate [7] (reprinted with permission of the Society for Experimental Stress Analysis).

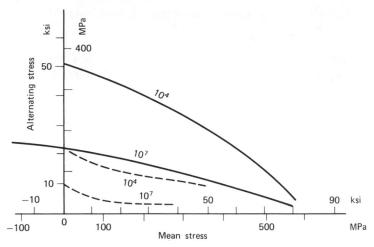

FIGURE 6.10 Constant life diagram for 7075-T6 wrought aluminum alloy with S_u = 570 MPa (82 ksi) (after Fig. 5.13) [3]. (——) Unnotched, (- - -) notched, K_t = 3.4.

reasonable limits we can predict K_f by the methods explained in the following sections.

6.3 THE BASELINE NOTCH FACTOR K_b

As a base for estimating the effect of other parameters we estimate the fatigue notch factor for zero mean stress and long life (10^7 cycles). We call it K_b. The difference between K_t and K_b is believed to be related to the stress gradient. Neuber [8] has proposed that the stress should be averaged over a small volume of material of depth a and has developed the following approximate formula for the notch factor.

$$K_b = 1 + \frac{K_t - 1}{1 + \sqrt{a/r}} \tag{6.7}$$

where r is the radius at the notch root. The length a depends on the material. Some suitable values are given in Table 6.3 [3]. Peterson* [6] has observed that good approximations can also be obtained by using the somewhat

* Rudolph E. Peterson (1901). Peterson received his B.S. and M.S. from the University of Illinois in 1925 and 1926, respectively. He joined Westinghouse Electric Corp. in Pittsburg in 1926 where he became manager of the Mechanics Department. He was very active in leadership and research roles in ASME, ASTM, and SESA and received many awards for his contributions. He was Chairman of the ASTM committee E-9 on fatigue, Chairman of the ASME applied mechanics division, and President of SESA.

TABLE 6.3   Neuber Lengths for Different Materials [3]

		Steel	
S_u, MPa (ksi)	500 (72)	1000 (145)	2000 (290)
a, mm (mils)^a	0.25 (10)	0.08 (3)	0.0002 (0.08)
		Aluminum	
S_u, MPa (ksi)	150 (22)	300 (43)	600 (87)
a, mm (mils)^a	2 (80)	0.6 (25)	0.4 (15)

^{a} 1 mil = 0.001 in.

similar formula

$$K_b = 1 + \frac{K_t - 1}{1 + p/r} \tag{6.8}$$

where p is another characteristic length. Values for p are given by Peterson [6], for instance, for steel of different strength levels:

S_u, MPa (ksi)	345 (50)	1725 (250)
Length p, mm (mils)	10 (400)	0.03 (1.3)

Studies of the correlation of the length a with test data were published by Kuhn and Hardrath [9] and by Kuhn and Figge [10].

These formulas and the associated characteristic lengths express the fact that for large notches with large radii we must expect K_b to be almost equal to K_t, but for small sharp notches we may find $K_b << K_t$ (little notch effect) for soft steel or soft aluminum, although K_b remains large for hard steel or hard aluminum. It has been said that hard metals are more notch sensitive than softer metals.

Looking at a numerical example we calculate K_b (according to Peterson's formula) for a 50 mm hole and for a 2 mm hole, in mild steel with $S_u = 345$ MPa and in hard steel with $S_u = 1725$ MPa. In all cases $K_t = 2.5$. We find the following values for K_b:

	S_u, MPa	
	345	1725
50 mm diameter	$K_b = 2.07$	$K_b = 2.50 = K_t$
2 mm diameter	$K_b = 1.14$	$K_b = 2.46$

6.4   EFFECTS OF STRESS LEVEL ON NOTCH FACTOR

For fatigue life of 10^6 to 10^7 cycles we can estimate the notched fatigue strength as S_f/K_b. The nominal stress calculated for these conditions is less than the yield strength. We can estimate the notch factor K_b as shown in Section 6.3.

In monotonic testing, notches may increase or decrease the nominal strength. A sharp groove in a notched tensile bar of ductile metal produces a higher nominal ultimate tensile strength than a uniform bar of the same minimum diameter. The smooth bar necks down and finally fails with a much reduced area. The notched bar cannot neck down because of the unyielded metal above and below the notch. Its ultimate tensile strength is greater. Data on the behavior of notched parts and smooth parts at different stress levels generally show that the notched fatigue strength is a greater fraction of the smooth fatigue strength as the nominal stresses get higher. Where data are available, the designer will of course use them, but in the absence of data the behavior of notched parts must be estimated.

We have seen in the previous section how one can estimate K_b, the base line notch factor, for fully reversed stresses at long life. In the absence of other data one can estimate the monotonic tensile strength of the notched part to be equal to the strength of the smooth part in monotonic testing, which corresponds, strictly speaking, to one-quarter of a fully reversed cycle but may be assumed to be 1 cycle. If a monotonic test can be made one would of course use its result rather than the estimate.

Two end points of the notched S–N curve are thus estimated, one at 1 cycle and the other at 10^7 cycles. A straight line between these points in a log S–log N plot is the best we can do unless other data are available. If the S–N line for smooth parts shows a pronounced curvature we would expect the curvature to appear also on the line for the notched part. An S–N curve for fully reversed stresses for notched parts in the usual laboratory environment can thus be estimated from the following data:

The ultimate tensile strength, S_u, of the material.

Its long life fully reversed fatigue strength, S_f, for smooth specimens of comparable size.

Its characteristic length a or p.

The elastic stress concentration factor, K_t, of the notch.

For example, for an 80 mm wide sheet of 1020 hot-rolled steel with a 10 mm central hole, we would construct the S–N curve in the following way. Fatigue strength at approximately 10^7 cycles and ultimate tensile strength are taken from Table A.1. They are 240 and 450 MPa (35 and 65 ksi), respectively. They define the upper line on Fig. 6.11, from 450 MPa at 1 cycle to 240 MPa at 10^7 cycles.

The elastic stress concentration fractor is 2.7 from Fig. 6.3. The characteristic length a is 0.25 mm (from Section 6.3). The base line notch factor then is

$$K_b = 1 + \frac{(2.7 - 1)}{(1 + \sqrt{a/r})} = 1 + \frac{1.7}{1.22} = 2.4$$

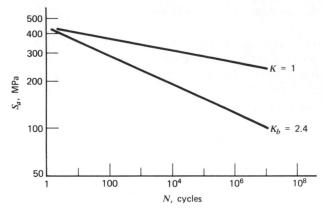

FIGURE 6.11 S–N diagrams for 1020 hot-rolled sheet steel with and without a notch.

The S–N line for the sheet with the hole then goes from 450 MPa at 1 cycle to 240/2.4 = 100 MPa at 10^7 cycles, as shown in Fig. 6.11.

6.5 MEAN STRESS EFFECTS ON NOTCHED PARTS

We know that stress range or strain range is the most important parameter of the fatigue life of smooth specimens. For notched parts this is true only if the local maximum stress is tensile and above a small threshold, above about 30 MPa (4 ksi). If there is no tensile stress there will be no fatigue failure. Can this be true for notched parts but false for smooth parts? No, but cycling between a maximum and minimum compressive stress is a fairly common regime for notched parts; in smooth specimens such a regime would usually produce buckling or yielding if the alternating stress is larger than the fully reversed fatigue strength.

Figure 6.12 shows lines for 10^6 cycles median fatigue life for smooth parts and for notched parts with $K_b = 2.9$ for the aluminum alloy 7075-T6. They are plotted in terms of alternating stress $S_a = (S_{max} - S_{min})/2$ versus mean stress $S_m = (S_{max} + S_{min})/2$. Note the different character of the line for notched parts as compared to the line for smooth parts and the great variation in the ratio K_f of their fatigue strengths: K_f is 2.9 at zero mean stress, increases to a maximum of 4.8 at 110 MPa mean stress, and decreases slowly at greater tensile mean stress. On the compression side it decreases rapidly to 1 at 165 MPa compressive mean stress and then becomes less than 1 at greater compressive mean stress, which means that a part with a groove may be stronger than a smooth part with the same minimum cross section. Can this be true? Yes, and we explain why in the next chapter. For now we observe that the fatigue lines end when the maximum or minimum stress is beyond the yield strength and explain how the diagram was constructed.

The fatigue strengths at zero mean stress were taken from Figure C-12 of Grover [3]. These data apply to a specimen with an elastic stress concentration factor 3.4. The slope of the line for smooth specimens was taken from the same figure. It might have been constructed by drawing the line to the point on the abscissa where S_m equals the true fracture stress.

The line for notched specimens consists of three segments: on the right side it is shown as $S = $ constant $= 30$ MPa, corresponding to the critical alternating tensile stress, S_{cat}, which is related, somewhat loosely, to the threshold range of the stress intensity factor. On the left side it is shown as $S_a = S_m + 60$ MPa, corresponding to the same alternating tensile stress of 30 MPa. In the center it passes through the given point $S_a = 63$ MPa $(K_b = 2.9)$ and has the same slope as the line for smooth specimens, because both S_a and S_m are increased in the same ratio (2.9) in the small volume near the root of the notch.

The yield strengths, 470 MPa monotonic on S_m and 525 MPa cyclic on S_a, were taken from the *SAE Handbook* [11] and joined by straight lines. The line constructed thus for notched parts conforms with the line obtained from experiments shown by Grover [3] except in the region of compressive mean stress, where no experimental line was shown. The equality of notched and smooth strengths in the presence of compressive mean stress conforms with general experience, for instance, with the data shown in Fig. 6.1. The area above the line N–N corresponds to nonpropagating cracks, which have often been observed in sharply notched parts.

The diagram in Fig. 6.12 is typical. Similar diagrams can easily be con-

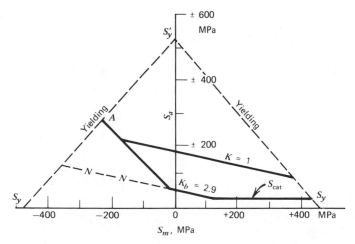

FIGURE 6.12 Haigh diagram for 7075-T6 aluminum alloy at 1 million cycles, with and without a notch.

structed for other materials and other values of K_b. The important points
to remember are:

Mean stress has more effect in notched parts than in smooth specimens.
Tensile mean stress can increase the notch factor K_f above the stress
concentration factor K_t. Tensile mean stress can be fatal in fatigue loading.
Compressive mean stress can wipe out the effects of stress concentrations
and save parts.

We see in the next chapter that mean stresses inherent in the unloaded part
are often much greater than the mean stresses caused by external loads.

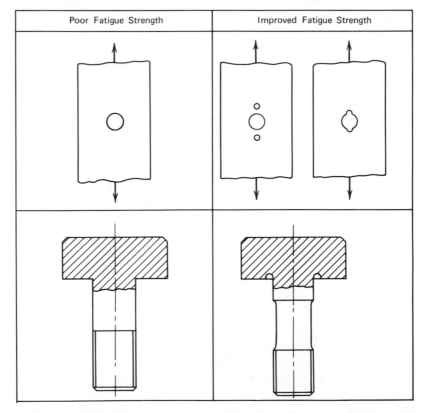

FIGURE 6.13   Mitigation of notch effects by design [12] (reprinted with permission
of Melbourne University Press).

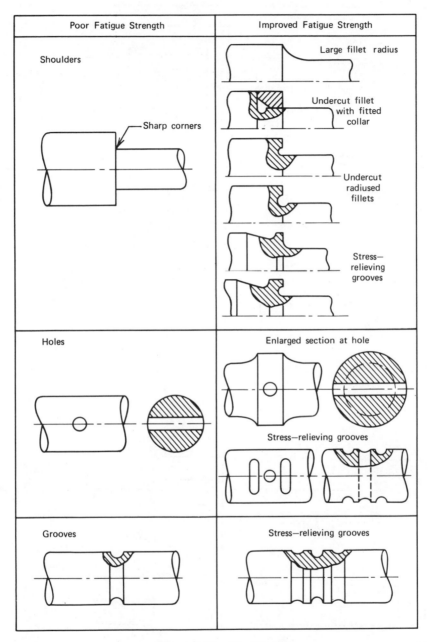

Poor Fatigue Strength	Improved Fatigue Strength

Shoulders

Large fillet radius

Sharp corners

Undercut fillet with fitted collar

Undercut radiused fillets

Stress—relieving grooves

Holes

Enlarged section at hole

Stress—relieving grooves

Grooves

Stress—relieving grooves

Figure 6.13 *(Continued)*

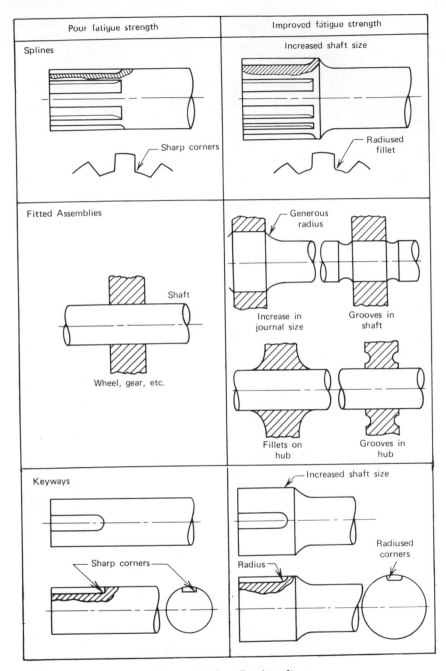

Figure 6.13 *(Continued)*

6.6 DOS AND DON'TS IN DESIGN

1. Do provide for smooth flow of stresses and forces. Figure 6.13 shows examples [12].
2. Do remember that parts may be strengthened by removing material in relief grooves and so on.
3. Don't permit sharp scratches or grooves with radii smaller than 0.25 mm (0.010 in.) on hard metals unless you know that they will not be subject to tensile stresses.
4. Don't worry about the exact numerical value of elastic stress concentration factors, but be aware of their trends.
5. Do consider design improvements like those shown in Fig. 6.13.

REFERENCES FOR CHAPTER 6

1. W. J. Harris, *Metallic Fatigue,* Pergamon Press, London, 1961, p. 48.
2. G. Kirsch, "Die Theorie Der Elastizitaet und Die Beduerfnisse Der Festigke-itslehre," *ZVDI,* 1898, mentioned by S. Timoshenko in *History of Strength of Materials,* McGraw-Hill Book Co., New York, 1953.
3. H. J. Grover, *Fatigue of Aircraft Structures,* NAVAIR 01-1A-13 Department of the Navy, 1966.
4. A. Buch, *Fatigue Strength Calculation Methods,* Technion (Israel) TAE report No. 314, Sept. 1977.
5. M. Creagar, "The Elastic Stress-Field Near the Tip of a Blunt Notch," Ph.D. Thesis, Lehigh University, Bethleham, PA, 1966.
6. R. E. Peterson, *Stress Concentration Factors,* John Wiley and Sons, New York, 1974.
7. P. E. Erickson and W. F. Riley, "Minimizing Stress Concentrations Around Circular Holes in Uniaxially Loaded Plates," *Exp. Mech.,* Vol. 18, No. 3, March, 1978, p. 97.
8. H. Neuber, *Kerbspannungslehre,* Springer, (Berlin), 1958; *Translation Theory of Notch Stresses,* U.S. Office of Technical Services, 1961.
9. P. Kuhn and H. F. Hardrath, "An Engineering Method for Estimating Notch-Size Effect in Fatigue," NACA TN 2805, 1952.
10. P. Kuhn and I. E. Figge, "Unified Notch Strength Analysis for Wrought Aluminum Alloys," NASA TN 1259, 1962.
11. "Technical Report on Fatigue Properties," SAE Information Report, J1099, *SAE Handbook,* 1978, p. 4.44.
12. J. Y. Mann, *Fatigue of Materials,* Melbourne University Press, Australia, 1967.

PROBLEMS FOR CHAPTER 6

1. Strain gages are often used to obtain strain–time histories and to obtain elastic stresses. If a strain gage is mounted in the y direction in Fig. 6.4 at a distance $x/r = 2.0$ (one radius away from the notch) and measured 0.005 strain for 7075-T6 aluminum, what is the stress S? How does this value compare with a uniaxial state of stress at a point obtained from $\sigma = E\epsilon$? What is the significance of the difference?

2. Derive the relationship between K_t based on gross area and K_t based on net area for the finite width plate of Fig. 6.3.

3. A stepped shaft, Figs. 6.5 to 6.7, has $D = 55$ mm, $d = 50$ mm, and $\rho = 5$ mm. Determine and compare K_t for (a) axial loading, (b) bending, and (c) torsion. Check other values in Figs. 6.5 to 6.7 and see if a conclusion can be drawn as to the effect of type of loading on K_t for a given geometry.

4. A threaded bolt has a series of stress concentrations at each thread. Using the streamline analogy, how does K_t at the thread adjacent to the shank compare with K_t at intermediate threads?

5. For the stepped shaft of problem 3, estimate the base line notch factor K_b for axial loading for the following steels: (a) Man-Ten, (b) RQC-100, (c) 4340 with $S_u = 745$ MPa, (d) 4340 with $S_u = 1260$ MPa, (e) 4340 with $S_u = 1530$ MPa, (f)4340 with $S_u = 1950$ MPa. Comment on the notch sensitivity of the six materials.

6. How can the fatigue strength for the shaft geometry of problem 3 be improved? List as many ways as you can think of to make this improvement.

7. Calculate K_f for the unpeened material of Fig. 6.1. Comment on the notch sensitivity of this unpeened material.

8. With reference to K_t for an elliptical notch, how important are cracks aligned parallel to a uniaxial tensile stress?

9. Superimpose approximate fully reversed S–N curves for the notched specimen and the six materials of problem 5.

CHAPTER 7

SELF-STRESSES AND NOTCH STRAIN ANALYSIS

We see in Chapter 6 that the notch factor K_f can be about the same as, greater than, or much smaller than the elastic stress concentration factor K_t, depending on the mean stress $(S_{max} + S_{min})/2$. To avoid fatigue failures we should try to avoid tensile mean stress and should try to have compressive mean stress. We can often achieve this by using *self-stresses*.

Self-stresses have also been called self-equilibrating stresses because they are in equilibrium within a part, without any external load. They have been called residual stresses because they may be left over from a previous operation.

Self-stresses exist in most manufactured parts. Their power to improve or ruin components subjected to millions of load cycles can hardly be overestimated.

7.1 AN EXAMPLE

Table 7.1 shows the effect of self-stresses produced by prestretching on the fatigue strength and fatigue notch factor of specimens of 4340 steel with two different notches heat treated to two different hardness levels [1].

The data of Table 7.1 compare the fatigue strengths at 2×10^6 cycles in rotating bending of round specimens with V-grooves. For the smooth specimens ($K_t = 1$) the fatigue strengths were 400 MPa (58 ksi) for the steel of 900 MPa ultimate tensile strength and 630 MPa (91 ksi) for that of 1700 MPa ultimate tensile strength. Stretching increased the fatigue strength of the specimens made of the softer steel by 130 percent for the sharper notch ($K_t = 3.2$) and by 90 percent for the more rounded notch ($K_t = 2.15$). The increases for the hard steel specimens were 220 and 155 percent. The damaging effect of the notches without stretching was greater for the hard steel. The self-stresses eliminated the notch effect almost completely. The worst parts were those made of the softer steel, with the sharp notch. The best parts were those made of the harder steel with the more rounded notch.

TABLE 7.1 Fatigue Strengths and Fatigue Notch Factors With and Without Self-Stresses

S_u, MPa		900			1700	
(K_t)	(1)	(2.15)	(3.2)	(1)	(2.15)	(3.2)
Without stretching						
S_f, MPa	400	205	160	630	240	190
(K_f)	(1)	(1.95)	(2.5)	(1)	(2.6)	(3.3)
With stretching						
S_f MPa	390	390	370	635	620	610
(K_f)	(1.03)	(1.03)	(1.08)	(0.99)	(1.02)	(1.03)

It is interesting to note that, with the self-stresses induced by stretching, the worst parts were much stronger than the best parts were without self-stresses. One can also see that the effects were not caused by work hardening the steel. The smooth specimens were essentially unaffected by the stretching operation, which had such a great influence on the fatigue strength of the notched parts.

To understand how stretching produces self-stresses and how the self-stresses improve the fatigue strength of the parts, it is necessary to do a notch strain analysis as explained in Section 7.5. Methods other than stretching are commonly used to produce self-stresses. They are briefly reviewed in Section 7.2.

7.2 PRODUCTION OF SELF-STRESSES

The many practical methods of producing beneficial compressive self-stresses on the surfaces of parts can be divided into two main groups: mechanical treatments and thermal treatments.

7.2.1 Mechanical Methods

The mechanical methods rely on strain gradients to produce local tensile yielding. The "springback" powered by the unyielded elastic metal in the part then produces compressive stresses in the stretched (tensile yielded) regions.

Figure 7.1 shows the stresses in a plate after cold bending. If stress concentrations are present, tensile overloading will do the job. Gerber's tests [2] show details of such treatment. The autofrettage of gun barrels [3] is another example. The tensile stresses in the thick walled tube are higher near the bore than on the outside. One can yield the bore by internal pressure without yielding the entire tube wall. When the pressure is removed the elastic contraction of the tube diameter produces compression in the metal near the bore. Overloading also is useful when cracks are present or suspected. If the cracks are small the overload retards their growth (reference 34 of Chapter 5). If the cracks are too large the overload serves as a proof test by breaking parts that are weakened too much.

The most common mechanical processes for producing beneficial self-stresses are shot-peening and surface rolling. Both use local plastic deformation, one by the pressure of the impact of small balls, the other by the pressure of narrow rolls. Surface rolling is widely used in the production of threads. It is very economical as a forming operation for bolts and screws, as well as beneficial for fatigue resistance. Heywood [4] has reported 50 percent greater fatigue strength for rolled threads as compared to cut or ground threads made of high strength steel. For steels of lower hardness the improvement is less. For less than 550 MPa (80 ksi) ultimate tensile strength there is no improvement according to Faires [5].

For internal threads, rolling is more expensive than cutting, but, because of the improved resistance against fatigue and corrosion fatigue, rolling is standard procedure for some sucker–rod couplings. Bellow and Faulkner [6, 7], obtained the results shown in Table 7.2 with AISI 8635 steel, quenched and tempered to Rockwell $C16$ to $C23$.

Bellow and Faulkner report that a shallow layer of compressive self-stress, produced by burnishing previously cut threads, improved the fatigue strength in air but slightly decreased the fatigue strength in salt water. The depth of the compressed layer is important.

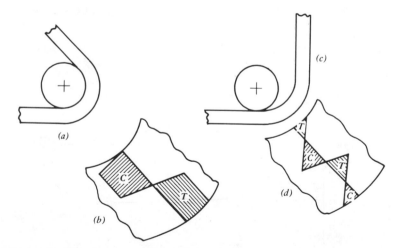

FIGURE 7.1 Self-stresses produced by bending a plate. (*a*) During bending. (*b*) Pattern of stresses assuming elastic–perfectly plastic behavior. (*c*) After release and springback. (*d*) Pattern of stresses after springback (T = tension stress, C = compression stress.

TABLE 7.2   Fatigue Strength in MPa (ksi) at 10^5 cycles, 0.2 Hz

Environment	Rolled Threads	Cut Threads
Air	510 (74)	303 (44)
3.5 percent NaCl	414 (60)	290 (42)
H_2S + CH_3COOH + 5 percent NaCl	317 (46)	<276 (40)

Rolling is also used for producing the desired self-stresses in the fillets between shank and head of bolts and for other fillets.

The most widely used mechanical process for producing self-stresses is shot-peening. Small balls (shot) are thrown at high velocities against the work pieces. They produce dimples and would produce considerable plastic stretching of the skin of the part if this were not restrained by the elastic core. Compressive stresses are thus produced in the skin. The depth of the compressive layer is determined by the material of the work piece and by the intensity of peening, which depends on size, material, and velocity of the shot. The magnitude of the compressive stress depends mainly on the material of the work piece. The intensity is specified in Almen numbers [8, 9]. Excessive intensities may produce excessive tensile stresses in the core of the work piece. Insufficient intensities may fail to provide enough protection. A chart of intensities that have been used successfully on a variety of parts has been published [10].

A typical stress distribution, produced by shot-peening, is shown in Fig. 7.2, the relation of the stress peak to material hardness obtained by Brodrick [11], is shown in Fig. 7.3, and the depth of compressive stress expected in steel and titanium as a function of peening intensity can be seen in Fig. 7.4

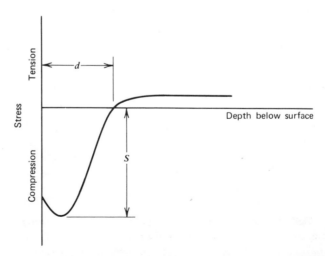

FIGURE 7.2   Typical distribution of self-stress below a shot-peened surface.

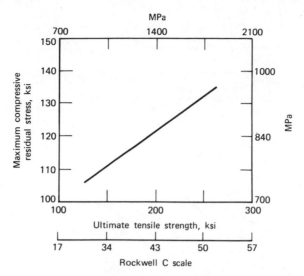

FIGURE 7.3 Residual stress produced by shot-peening versus tensile strength of steel [11].

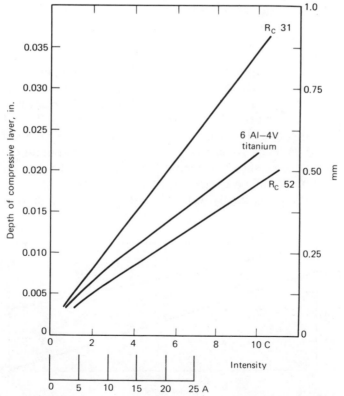

FIGURE 7.4 Depth of compressive stress versus Almen intensity for steel and titanium [12] (reprinted with permission of the Metal Improvement Co.).

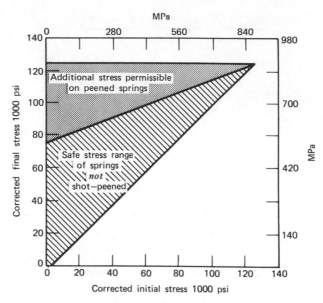

FIGURE 7.5   Safe-stress ranges for coil springs (0.207 diameter steel) [13] (reprinted with permission of the Associated Spring Corp.).

[12]. Shot-peening is used on many parts, from small blades for chain saws to large crankshafts for Diesel locomotives. Application to high performance gears and to springs is almost universal. Figure 7.5 shows recommended working stresses for peened and unpeened valve springs [13]. It indicates that for peened springs the working stresses are no longer limited by fatigue failures but by yielding, which is less insidious and seldom fatal. Figure 7.6, for carburized gears, shows a tenfold life increase and a strength increase of 40 percent at 10 million cycles as obtained by Straub [14].

Self-stresses are especially valuable when used with harder materials

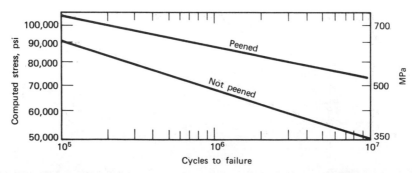

FIGURE 7.6   Fatigue chart of carburized gears [14] (reprinted with permission of McGraw-Hill Book Co.).

because the full potential of greater yield strength can be used only if the damaging effect of notches and scratches can be overcome. Compressive self-stresses can do just that, as indicated in Fig. 7.7 from data by Brodrick [11]. Adequate depth of the compressively stressed layer is important, because cracks propagate about as fast in harder as in softer material, and the critical crack length is less. The compressed layer must be deep enough to be able to stop the cracks.

Other mechanical processes that achieve improvement of fatigue strength by compressive self-stresses include coining around holes, expansion of holes, and hammer peening of welds. Several references are listed at the end of this chapter [15–18].

7.2.2. Thermal Processes

Thermal processes are perhaps the oldest means of improving the fatigue resistance of parts. Surface hardening of steel is the chief example. If it is properly done it leaves parts with a skin that is hard and in compression. The compressive stress can reach the yield strength of the hardened skin; it very effectively prevents the growth of cracks and thus permits us to realize the gain in fatigue strength that we would expect from the increased hardness, as shown in Fig. 7.7.

Surface hardening can be accomplished by carburizing, nitriding, induc-

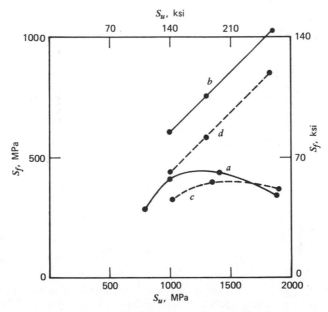

FIGURE 7.7   Fatigue limit as a function of ultimate tensile strength for peened and unpeened specimens [11]. (*a*) Shaft, not peened, (*b*) shaft, peened, (*c*) scratched plate, not peened, (*d*) scratched plate peened.

tion hardening, severe quenching of carbon steel, or similar methods. Figure 7.8 shows the profile of compressive self-stresses produced by carburizing and the effect of tempering in modifying the profile [19]. Such data are important for matching heat treatment with loads and stresses [20]. Figure 7.9 shows the fatigue limit of several shallow quenched steels versus the

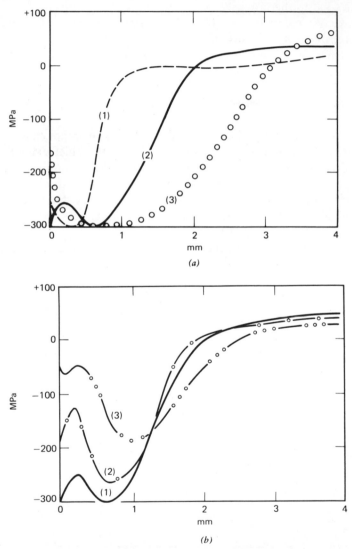

FIGURE 7.8 Self-stresses in carburized steel bars, 19 mm diameter, plotted versus depth below surface (8617 steel) [19] (reprinted with permission of McGraw-Hill Book Co.). (*a*) Effect of case depth (1) 0.7, (2) 1.3 and (3) 2.2 mm. (*b*) Effect of tempering on 1.3 mm original case depth, (1) as quenched, (2) tempered at 150°C (300°F), and at (3) 190°C (375°F).

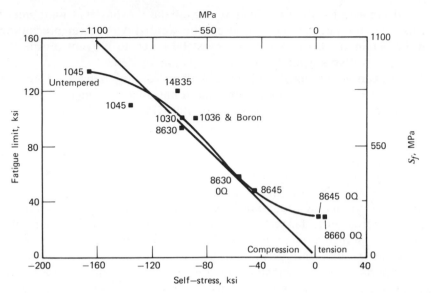

FIGURE 7.9 Effect of self-stress on fatigue limit of notched steel bars [21, 22]. (Data taken with permission of Society of Automotive Engineers). OQ, oil quenched; all others, water quenched.

measured compressive self-stresses [21, 22]. The fatigue limit is almost exactly equal to the measured compressive self-stress. The depths of the compressed layers are adequate for arresting cracks.

Thermal treatments can also produce the opposite effect. The heat applied in welding can produce tensile stresses up to the yield strength of the material. They reduce fatigue strength and exacerbate the effects of notches and cracks. Severe grinding also produces heat that leads to tensile stresses in the surface and decreases fatigue strength to less than half of its expected value. Many examples of the effects of residual stresses (self-stresses) are given in the book by Almen and Black [23]. Others are shown in reference 24, which contains 13 chapters on Fatigue Considerations Resulting from Processing, each with a bibliography. Beside these there is a chapter by Horger, on Residual Stresses, with a bibliography of 177 items.

7.3 PERMANENCE OF SELF-STRESSES

Self-stresses produce the same effects that mean load stresses of the same amount and distribution would produce, but there is a difference. The mean load stresses persist as long as the mean load remains. The self-stresses persist as long as the sum of self-stress and applied load stress does not exceed the yield strength of the materials. Thus they are more beneficial (and potentially more harmful) when applied to hard metals with high yield strengths. In softer metals such as mild steel the self-stresses can be more easily decreased by yielding. This is one of the reasons that mild steel is not usually shot-peened and why it can be welded with fewer precautions than harder materials. Greater care must be taken to prevent tensile "residual" self-stresses when harder materials are welded.

As an example we may think of a shaft in rotating bending. Using approximate round numbers we assume that it has a yield strength of 700 MPa (100 ksi) and compressive self-stresses of 400 MPa (60 ksi) in the skin layer. Neglecting strain hardening we can say that an applied alternating stress of 550 MPa (80 ksi) will reduce the self-stress to 150 MPa (20 ksi), because the maximum compressive stress cannot be 400 + 550 = 950 MPa, but only 700 MPa. Removing the applied stress leaves self-stress of 700 − 550 = 150 MPa (100 − 80 = 20 ksi).

In constant amplitude long-life fatigue, when millions of cycles must be endured, there is little danger or hope of decreasing the beneficial or harmful self-stresses in high strength materials. In the intermediate life range we get intermediate results. In practice shot-peening is very successfully used not only for parts that must withstand millions of cycles, such as valve springs, but also on parts such as axle shafts, which are expected to last a few hundred thousand cycles of maximum stress range, or landing gears, which must withstand only a few thousand cycles of maximum stress and many cycles of smaller stress range.

Note that loading in one direction only, as in springs and most gears, will not destroy beneficial self-stresses. Automobile leaf springs for instance are usually shot-peened. The load stress is tensile on the upper side. If the sum of self-stress and applied stress is always less than the yield strength, it will not produce yielding and will not decrease the self-stress on the peened side of the leaf.

On the lower side of the leaf the load stress is compressive. If that side were peened the load stress and self-stress together would produce yielding. But fatigue failures will not develop on the compressively stressed side of the leaf. Therefore, it would not matter if the peening stresses would diminish there. As a matter of fact, only the tension side is peened because tests have shown that peening the compressive side would be an unnecessary expense.

In springs, as in other parts that are loaded predominantly in one direction, an overload applied early in the life is beneficial. Springs, hoists, and pressure vessels are strengthened by proof loading with a load higher than the highest expected service load.

7.4 MEASUREMENT OF SELF-STRESSES

Because self-stresses are so important it is desirable to verify their existence by measurement. This can be done fairly easily as far as stresses on the surface are concerned, but for measurement of the depth to which the stresses extend no practical nondestructive method has yet been developed.

Measurement of surface stresses is done by X-ray diffraction methods [25]. Automated X-ray machines for this purpose are commercially available. Measurement in depth still requires removal of material.

One can remove surface layers and repeat X-ray measurements on the newly exposed surface, or one can drill small holes surrounded by strain gages attached to the surface. The changes in strain readings as the hole is made deeper permit approximate calculation of the stress distribution [26, 27]. Measurement of self-stresses is still the exception rather than the rule. Most users of self-stresses rely on careful control of processes such as carburizing, shot-peening, and coining.

7.5 NOTCH STRAIN ANALYSIS

The loads on notched parts are often so high that the local stress calculated from nominal stress and notch factor by the formula $\sigma = K_t S$ is considerably above the yield strength. This is especially true for parts designed for only a few thousand load cycles, but may be true for others too. When the local stress exceeds the yield strength it will be less than $K_t S$ and we can no longer use the formula $\sigma = K_t S$. We use instead a formula for strain

$$\epsilon = K_\epsilon \dot{e}$$

where ϵ = local strain at a notch
           e = nominal strain
           K_ϵ = strain concentration factor

We can then estimate the fatigue life to small cracks from a strain–life curve or obtain the stress corresponding to ϵ from a stress–strain curve and use it to estimate fatigue life from a stress versus life curve.

Note that uniformly repeated load cycles impose uniformly repeated strain cycles on the metal at the root of the notch, as long as most of the part remains elastic. If the metal at the notch root is strained beyond the yield strength it may strain harden and cyclically harden or soften as is discussed in Chapter 3. The material at the notch root is under "strain control."

We are now primarily interested in the strain range $\Delta\epsilon$ at the notch root. As long as stresses and strains at the notch root remain elastic we have

$$\sigma = K_t S \qquad \epsilon = K_t e \tag{7.1}$$

When $K_t S$ becomes greater than S_y the material at the notch root yields; stresses then are no longer proportional to strains. We then define

$$K_\epsilon = \frac{\epsilon}{e} \tag{7.2}$$

$$K_\sigma = \frac{\sigma}{S} \tag{7.3}$$

The relation between σ and ϵ is given by the stress–strain curve, which is represented by

$$\epsilon = \epsilon_e + \epsilon_p = \frac{\sigma}{E} + \left(\frac{\sigma}{K}\right)^{1/n} \tag{7.4}$$

Values for n and K can be taken from Table A.2.

The question to be resolved by notch-strain analysis is, given the nominal elastic stress S or strain e, what are the local stress σ and the local strain ϵ at the notch root surface and near the surface?

Answers to this question can be obtained by finite element methods and for some cases by the equations of the theory of plasticity. For most fatigue problems, however, one is satisfied with estimating the conditions at the surface of the notch root by considering two extreme cases. The first case is:

$$\frac{\epsilon}{e} = K_\epsilon = K_t \tag{7.5}$$

The strain can be computed directly and, if desired, the stress can then be obtained from the stress–strain curve. Equation 7.5 expresses the linear rule or strain concentration invariance. It agrees with measurements in plane strain situations, such as circumferential grooves in shafts in tension or bending.

The second case is:

$$\epsilon = \frac{K_t^2}{K_\sigma} e \qquad K_\epsilon K_\sigma = K_t^2 \tag{7.6}$$

This is known as Neuber's rule because he derived it for longitudinallly grooved shafts in torsion [28]. It agrees with measurements in plane stress situations, such as thin sheets in tension. Application of this rule requires the solution of two simultaneous equations:

$$\epsilon\sigma = K_t^2 e S \tag{7.7}$$

$$\epsilon = \epsilon_e + \epsilon_p = \frac{\sigma}{E} + \left(\frac{\sigma}{K}\right)^{1/n} \tag{7.4}$$

On σ–ϵ coordinates, Eq. 7.7 describes a hyperbola and Eq. 7.4 describes the stress–strain curve. The intersection of these two curves defines the desired values σ and ϵ

Figure 7.10 shows the application, by graphic methods, of both these rules to a notch with $K_t = 3$ in a hypothetical material with the stress–strain curve

$$\epsilon = \frac{\sigma}{60{,}000} + \left(\frac{\sigma}{2000}\right)^8$$

which is Eq. 7.4 with $E = 60{,}000$ MPa, $K = 2000$ MPa, and $n = 0.125$.

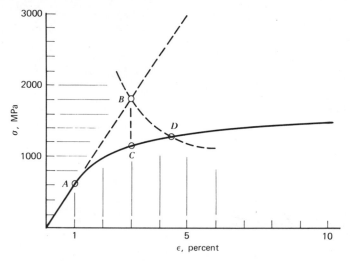

FIGURE 7.10   Notch strain determination by the linear rule and by Neuber's rule.

For a nominal stress of 600 MPa:

Point A represents the nominal conditions $S = 600$ MPa, $e = 1$ percent.
Point B represents the fictitious elastic stresses and strains $\sigma = 3 \times 600 = 1800$ MPa, $\epsilon = 3 \times 1$ percent $= 3$ percent.
Point C is the solution by the linear rule: $\epsilon = 3$ percent, $\sigma = 1130$ MPa.
Point D is the solution by Neuber's rule: $\epsilon = 4.4$ percent, $\sigma = 1230$ MPa.

Note that $\epsilon \times \sigma = 4.4$ percent $\times 1230$ MPa $= 54$ MPa is equal to $3e \times 3S = 9 \times 1$ percent $\times 600$ MPa $= 54$ MPa.

The discussion of notch strain analysis up to this point applies to mon-otonic loading. For cyclic loading the monotonic stress–strain curve is replaced by the hysteresis curve and the strains and stresses are replaced by the strain ranges and stress ranges, as shown in Fig. 7.11. Note that the hysteresis curve is geometrically similar to the cyclic stress–strain curve, but twice as large, and that the origin of the Neuber hyperbola is at the point of reversal, not at $\sigma = 0$. From the two reversal points, the hysteresis loop is drawn as a straight line for stress change of $2S_y'$. The two rules for computing $\Delta\epsilon$ then are:

$$\text{Linear rule,} \qquad \Delta\epsilon = K_t \, \Delta e \qquad\qquad (7.7)$$

$$\text{Neuber's rule,} \qquad \Delta\epsilon \, \Delta\sigma = K_t^2 \, \Delta e \Delta S \qquad\qquad (7.8)$$

The example in Section 7.7 shows the application of Neuber's rule in cyclic loading.

Figure 7.12 shows the results of strain range measurements in cyclically loaded notched specimens by Kotani et al. [29]. The ratio $\Delta\epsilon/(K_t\Delta e)$ is plotted over the ratio $(K_t\Delta S)/2S_y'$. Yielding starts when $(K_t \, \Delta S)/2S_y' = 1$.

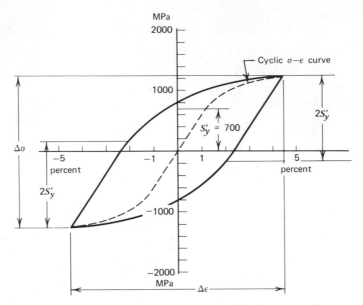

FIGURE 7.11   A cyclic stress–strain curve and a hysteresis loop.

In these specimens K_t was slightly more than 2. According to the linear
rule the ratio $\Delta\epsilon/K_t\,\Delta e$ should remain equal to one. This is substantially
true for the grooved shafts. According to Neuber's rule the curve for
$\Delta\epsilon/K_t\,\Delta e$ should turn sharply upward as yielding begins. This is true for
the thin sheet ($t = 6$ mm). The notch in the thicker plate ($t = 60$ mm) acts
in an intermediate manner, between $K_\epsilon = K_t^2/K_\sigma$ and $K_\epsilon = K_t$.

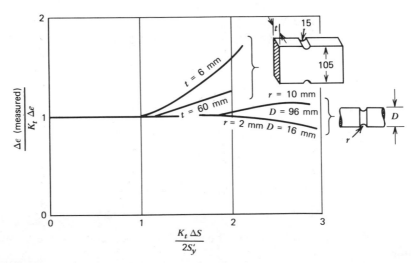

FIGURE 7.12   Local plastic strain in notches of medium carbon steel, cyclic push-
pull, gage length 1 mm [29].

If one would require an intermediate formula, it might be

$$\Delta\epsilon = K_t \, \Delta e \left(\frac{K_t}{K_\sigma}\right)^m \tag{7.9}$$

where $m = 1$ for plane stress, $m = 0$ for plane strain, and $0 < m < 1$ for intermediate situations.

Figure 7.12 thus shows that the Neuber assumption is good for thin sheets and that it is conservative by predicting strains higher than measured strains in almost all other cases.

By considering the elastic constraint at the notch we can estimate a suitable value for K_ϵ from Fig. 7.12. It is usually between K_t and k_t^2/K_σ.

7.6 EXAMPLE OF THE APPLICATION OF NOTCH STRAIN ANALYSIS WITH THE LINEAR RULE, $K_\epsilon = K_t$

A round shaft with a circumferential groove is subject to rotating bending stresses. It has a minimum diameter of 12.7 mm (0.5 in.), a groove radius of 0.38 mm (0.015 in.), and $K_t = 3.20$. The bending moment is 72.3 N.m (640 in. lb) resulting in a nominal alternating stress of 360 MPa (52.2 ksi). The material is 4340 steel, heat treated to Rockwell C24. Yield strength is 835 MPa (121 ksi). The tensile fracture load of such a shaft is 187,000 N (42,000 lb), corresponding to a nominal stress of 1580 MPa (229 ksi). Ultimate tensile strength of smooth tensile specimens is 900 MPa (130 ksi), which indicates substantial notch strengthening.

The local alternating stress at the notch root, calculated as $\sigma_a = K_t S_a = K_t(\Delta S/2) = 3.2 \times 360 = 1152$ MPa is well above the yield strength of 835 MPa. We can estimate the fatigue life from notch strain analysis. For the round shaft with groove we use $K_\epsilon = K_t$. The local alternating strain ϵ_a then is from Eq. (7.5):

$$\epsilon_a = K_t \frac{S_a}{E} = 3.2 \times \frac{360}{200,000} = 0.58 \text{ percent}$$

The strain–life relation for this steel is shown as a graph on log–log coordinates in Fig. 7.13. This graph was developed from the following values:

$$\sigma_f' = 1240 \text{ MPa} \qquad \left(\frac{\sigma_f'}{E} = 0.62 \text{ percent}\right)$$

$$\epsilon_f' = 0.6$$

$$b = -0.08$$

$$c = -0.6$$

interpolated from values given for 4340 steels in Table A.2.

From the graph we find $2N = 90,000$ or $N = 45,000$ cycles to failure, defined as the appearance of observable cracks. If the notch strain were calculated by Neuber's rule we would require a stress–strain curve. The strain range would be computed to be about 1 percent. With 1 percent the predicted life to cracking would be less than 1000 cycles.

The difference between "failure" and "breaking" is important. The part described in this example actually was unbroken after 4 million cycles because it had been prestretched. When pulled apart after fatigue testing it showed a crack 0.48 mm (0.019 in.) deep extending completely around the circumference at the root of the notch. The crack very likely had started after a few thousand cycles as estimated above. The crack could not propagate because it was arrested in a layer of material in which the compressive self-stress was greater than the applied tensile stress.

This example shows that total fatigue life cannot be predicted from analysis of load stresses alone. One must also consider the effect of the self-stresses.

In the next chapter we see that the nominal fatigue limit with compressive self-stresses of sufficient magnitude and depth corresponds approximately to a nominal stress equal to half the yield strength, regardless of notch factor (see point A, Fig. 6.12).

The bar that lasted over 4 million cycles had been prestretched with a load almost equal to the fracture load. This produced very deep yielding. On release of the prestretching load the bar contracted elastically. Near the notch root the contraction strain produced compressive self-stress, balanced by tensile self-stress in the interior of the bar. Figure 7.14 from Gerber [2] shows the distribution of the axial stress during prestretching and after release of the load. Note the yielding in compression after the release. The zone of yielded material is cross hatched. The stress distribution shown in Fig. 7.14 was calculated by methods of the theory of plasticity (slip line

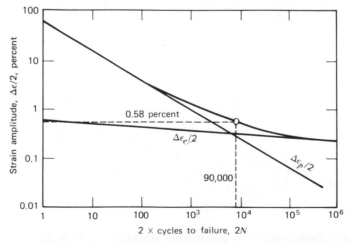

FIGURE 7.13 Prediction of fatigue cracking by notch strain analysis.

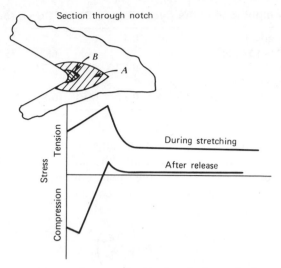

FIGURE 7.14 Yield zones and stress distribution during and after severe stretching [2]. (*A*) Zone yielded during stretching; (*B*) Zone yielded in compression on release of tensile load.

fields) and by matching axial and radial stresses at the plastic–elastic interface [2].

One can estimate the self-stresses from an approximate notch strain analysis. For instance, if the nominal stress caused by the preload is 500 MPa, the nominal strain is $e = 0.25$ percent and the strain at the notch root $\epsilon = K_t e = (3.2)(0.25 \text{ percent}) = 0.8$ percent. The monotonic stress–strain curve of this material is practically horizontal after yielding, up to 1 percent strain. The stress at the notch root, during prestretching, is the yield strength or 835 MPa (121 ksi). On release the strain decreases almost as much as it increased by loading, because most of the cross section remains elastic. A decrease of 0.8 percent, produces a stress decrease of $0.08E = 1600$ MPa. The compressive stress is then $835 - 1600 = -765$ MPa. This is the compressive self-stress at the surface. It may be less than the applied tensile stress. To determine whether cracks will be arrested one must make a notch strain analysis in depth, as in Fig. 7.14.

7.7 EXAMPLE OF NOTCH STRAIN ANALYSIS USING NEUBER'S RULE

A sheet of RQC-100 steel has a notch, $K_t = 3.2$. It is subject to fatigue loads that produce nominal alternating stresses ±360 MPa (52.2 ksi) on the net section. The cyclic yield strength, S_y', of this steel is 600 MPa. What are the alternating strains at the notch? The nominal strain range Δe is (nominal stress range/modulus of elasticity) $= \Delta S/E = 720 \text{ MPa}/200{,}000$

TABLE 7.3   Computation of the Cyclic Stress–Strain Curve for RQC-100 Steel

$\Delta\epsilon_p/2$ (percent) given	0	0.1	0.2	0.5	1	2	5
$\Delta\sigma/2$ MPa $= 1434\,(\Delta\epsilon_p/2)^{0.14}$	0	545	601	683	733	829	943
$\Delta\epsilon_e/2$ (percent)	0	0.27	0.30	0.34	0.38	0.41	0.47
$\quad = 1434\,(\Delta\epsilon_p/2)^{0.14}/2000$							
$\Delta\epsilon/2$ (percent) $= \Delta\epsilon_p/2 + \Delta\epsilon_e/2$	0	0.37	0.50	0.84	1.38	2.41	5.47

MPa $= 0.36$ percent. To find the strain range at the notch root we use the cyclic stress–strain curve because under the influence of fatigue loads the material will soon approach the stable cyclic condition. With the data from Table A.2 the cylic stress–strain curve can be computed by calculating $\Delta\sigma/2$ from

$$\frac{\Delta\sigma}{2} = K'\left(\frac{\Delta\epsilon_p}{2}\right)^{n'} \tag{3.3}$$

Then $\Delta\epsilon_e/2$ can be obtained from

$$\frac{\Delta\epsilon_e}{2} = \frac{\Delta\sigma}{2E} \tag{7.9}$$

and the total strain range can be calculated from

$$\Delta\epsilon = \Delta\epsilon_e + \Delta\epsilon_p \tag{3.1}$$

as shown in Table 7.3. This cyclic stress–strain curve is shown in Fig. 7.15. A plot of stress versus strain at the notch root forms a hysteresis loop (Fig. 7.16). The rising and falling stress–strain curves are assumed to be geometrically similar to the cyclic stress–strain curve but twice as large because the cyclic curve is a plot of $\Delta\sigma/2$ versus $\Delta\epsilon/2$. The hysteresis loop shows the notch root stress and strain ranges $\Delta\sigma$ and $\Delta\epsilon$. The rising part (segment D–A–B in Fig. 7.16) of the loop is shown as Fig. 7.17. Point X in Fig. 7.17 shows a strain range of 1.15 percent and a stress range of 2300 MPa. These would be the values if the material were elastic through these stress ranges

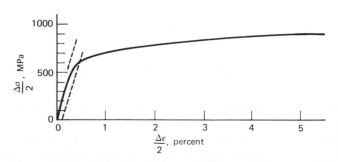

FIGURE 7.15   The cyclic stress–strain curve for RQC-100 steel.

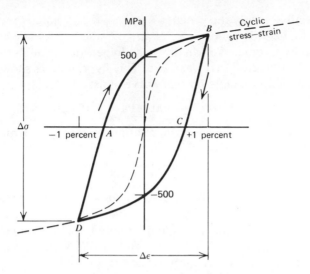

FIGURE 7.16 A hysteresis loop for 1 percent strain amplitude for RQC-100 steel.

and strain ranges. We would then have

$$\Delta\sigma = K_t\,\Delta S = 3.2 \times 720 = 2300 \text{ MPa}$$

$$\Delta\epsilon = K_t\,\Delta e = 3.2 \times 0.36 \text{ percent} = 1.15 \text{ percent} = \frac{\Delta\sigma}{E}$$

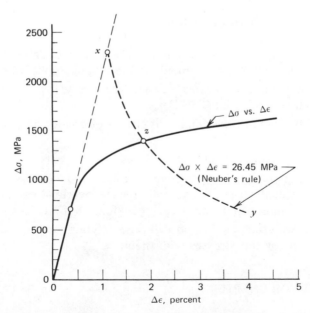

FIGURE 7.17 Determination of the strain at the root of a notch according to Neuber's rule; material: RQC-100; nominal stress range 720 MPa; $K_t = 3.2$.

According to Neuber's rule the product $K_t \Delta e \times K_t \Delta S$ equals the product $\Delta\epsilon \Delta\sigma$ because by his rule $K_\epsilon K_\sigma = K_t^2$. Pairs of values that have a constant product define a hyperbola. We can draw the hyperbola through the known point $\Delta\sigma = 2300$ MPa, $\Delta\epsilon = 1.15$ percent. This hyperbola is shown in Fig. 7.17 as the line X–Y. All points on that curve satisfy the relation

$$\Delta\sigma \, \Delta\epsilon = 26.45 \text{ MPa} = 2300 \text{ MPa} \times 1.15 \text{ percent}$$

Point Z also satisfies the stress–strain relation for the given material. It provides the solution

$$\Delta\sigma = 1400 \text{ MPa} \qquad \Delta\epsilon = 1.9 \text{ percent}$$

One can then predict a fatigue life either from an ϵ–N curve or from an S–N curve.

The computation of $\Delta\sigma$ and $\Delta\epsilon$ is not required for the application of Neuber's rule if the fatigue life is taken not from an S–N curve nor from an ϵ–N curve, but from a curve of the product eS plotted over N. For any notch factor K_b one need only find the life that corresponds to the product $K_b^2 eS$. This approach was proposed by Smith et al. [30, 31].

In this book we use the linear rule unless otherwise stated. We believe that it more closely describes the conditions in most machine parts and that it is easier to understand and to apply. Neuber's rule is more realistic for thin sheets and always more conservative than the linear rule. Additional examples are given in Chapter 8.

7.8 DOS AND DON'TS IN DESIGN

1. Do consider the beneficial and the harmful effects of self-stresses for all long-life (high cycle) and intermediate-life fatigue applications.
2. Don't expect much help from self-stress in very low cycle (less than 10,000 cycles) applications.
3. Do take advantage of the combined effects of greater hardness and skin compression.
4. Don't permit surface tensile self-stresses in hard parts.
5. Do remember that grinding and welding can produce very harmful self-stresses (residual stresses) and that tensile overloading, peening, and surface hardening can produce very beneficial self-stresses.
6. Do consult with the people who will process your parts and discuss your requirements for self-stresses with them.

REFERENCES FOR CHAPTER 7

1. H. O. Fuchs, "Regional Tensile Stress as a Measure of the Fatigue Strength of Notched Parts," *Mechanical Behavior of Materials*, Proceedings of the

International Conference, Vol. II, Kyoto, Society of Materials Science, Japan, 1972, p. 478.

2. T. L. Gerber, "The Effect of Tensile Preload on the Fatigue Strength of Notched Steel Bars," Ph.D. Dissertation, Stanford University, 1967. (Results reported in *Achievement of High Fatigue Resistance in Metals and Alloys,* ASTM STP 467, 1970, p. 276.

3. T. E. Davidson, D. P. Kendall, and A. N. Reiner, "Residual Stresses in Thick-Walled Cylinders Resulting from Mechanically Induced Overstrain," *Exp. Mech.,* Vol. 3, No. 11, Nov. 1963, p. 253.

4. R. B. Heywood, *Designing Against Fatigue of Metals,* Reinhold, New York, 1962, p. 272.

5. V. M. Faires, *Design of Machine Elements,* 4th ed., Macmillian, London, 1965, p. 162.

6. D. G. Bellow and M. G. Faulkner, "Development of an Improved Internal Thread for the Petroleum Industry," *Closed Loop,* MTS Systems Corp., Vol. 8, No. 1, Feb. 1978, p. 3.

7. D. G. Bellow and M. G. Faulkner, "Salt Water and Hydrogen Sulfide Corrosion Fatigue of Work-Hardened Threaded Elements," *J. Test. Eval.,* Vol. 4, No. 2, 1976, p. 141.

8. "Test Strip, Holder and Gage for Shot-Peening," SAE Standard J 442, *SAE Handbook,* 1979, p. 9.05.

9. H. O. Fuchs, "Shotpeening Effects and Specifications," *Metals,* ASTM STP 196, 1962, p. 22.

10. "Shot-Peening," *Metals Handbook,* Vol. 2, 8th ed., American Society of Metals, Metals Park, OH, Vol. 2 1964, Fig. 7, p. 398.

11. R. F. Brodrick, "Protective Shot-Peening of Propellers," Wright Air Development Center Technical Report TR55-56, 1955.

12. *Shot-Peening Applications,* 5th ed., Metal Improvement Co., 1977.

13. *Spring Design and Selection,* Associated Spring Corp., Bristol, Conn, 1956.

14. J. C. Straub, "Shot-Peening" *Metals Engineering, Design,* 2nd ed., O. J. Horger, Ed., McGraw-Hill Book Co., New York, 1965, p. 258.

15. "Surface Rolling and Other Methods for Mechanical Prestressing of Metals," Technical Report HS 3, Society of Automotive Engineers.

16. A. M. Nawwar, J. Shewchuk, and D. T. Lloyd, "The Improvement of Fatigue Strength by Edge Treatment, *Exp. Mech.,* Vol. 15, No. 5, May 1975, p. 161.

17. A. M. Nawwar and J. Shewchuk, "On the Measurement of Residual Stress Gradients in Aluminum Alloy Specimens," *Exp. Mech.,* Vol. 18, No. 7, July 1978, p. 260.

18. E. R. Speakman, "Fatigue Life Improvement Through Stress Coining Methods," *Achievement of High Fatigue Resistance in Metals and Alloys,* ASTM STP 467, 1970, p. 269.

19. R. L. Mattson and G. H. Robinson, "Case Carburizing" *Metals Engineering, Design,* 2nd ed., O. J. Horger, Ed., McGraw-Hill Book Co., New York, 1965, p. 288.

20. V. K. Sharma, G. H. Walter, and D. H. Breen, "Predicting Case Depths for Gears," *Prod. Eng.,* June 1979, p. 49.

21. D. V. Nelson, R. E. Ricklefs, and W. P. Evans, "The Role of Residual Stresses in Increasing Long Life Fatigue Strength of Notched Machine Members," *Achievement of High Fatigue Resistance Metals and Alloys,* ASTM STP 467, 1970, p. 228.

22. R. B. Liss, C. G. Massieon, A. S. McKloskey, "The Development of Heat Treat Stresses and Their Effect on Fatigue Strength of Hardened Steel," SAE paper No. 650517, 1965.

23. J. O. Almen and P. H. Black, *Residual Stresses and Fatigue in Metals,* McGraw-Hill Book Co., New York, 1963.

24. O. J. Horger, Ed., *Metals Engineering, Design,* 2nd ed., McGraw-Hill Book Co., New York, 1965.

25. "Residual Stress Measurement by X-Ray Diffraction," Technical Report HS J 784a Society of Automotive Engineers.

26. N. J. Rendler and I. Vigness, "Hole-drilling Strain-gage Method of Measuring Residual Stress," *Exp. Mech.,* Vol. 6, No. 12, Dec. 1966, p. 577.

27. J. P. Sandifer and G. E. Bowie, "Residual Stress by Blind-hole Method with Off-center Hole," *Exp. Mech.,* Vol. 18, No. 5, May 1978, p. 173.

28. H. Neuber, "Theory of Stress Concentration for Shear-Strained Prismatical Bodies with Arbitrary Nonlinear Stress Strain Laws," *Trans. ASME, J. Appl. Mech.,* Vol. 28, Dec. 1961, p. 544.

29. S. Kotani, K. Koibuchi, and K. Kasai, "The Effect of Notches on Cyclic Stress–Strain Behavior and Fatigue Crack Initiation," *Proc. 2nd Int. Conf. Mech. Behav. Mater., Boston, MA,* 1976, p. 606.

30. K. N. Smith, P. Watson, and T. H. Topper, "A Stress–Strain Function for the Fatigue of Metals," *J. Mater.,* Vol. 5, No. 4, Dec. 1970, p. 767.

31. K. N. Smith, M. El Haddad, and J. F. Martin, "Fatigue Life and Crack Propagation Analyses of Welded Components Containing Residual Stresses, *J. Test. Eval.,* Vol. 5, No. 4, 1977, p. 327.

PROBLEMS FOR CHAPTER 7

1. A flat specimen similar to, say, the Almen strip is shot-peened on one side as shown. Qualitatively sketch a proper stress distribution along section A–B, such that all equations of equilibrium are qualitatively satisfied. This assumes $\Sigma F = \Sigma M = 0$.

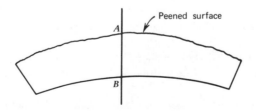

2. Each day 5000 heavy beer barrels are rolled down the sloping cantilever beam to point A, where a crane picks them up for shipment. About once a year the cantilever beam breaks. It has been suggested that compressive self-stresses might increase the life of the beam. Which way would you overload the beam so that surface compressive self-stresses would be formed at the proper location? Would this be a good solution if failure occurred once a week? Why?

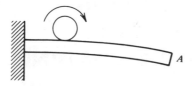

3. A 100 mm × 150 mm rectangular beam is subjected to pure bending as shown. $M = 1.4\ M_0$ where M_0 is the moment that just causes the first fiber to yield. Assume an ideally elastic–plastic material with the given stress–strain properties in both tension and compression and no Bauschinger effect. Determine the resultant self-stress distribution if the moment is completely removed. Make a final plot of the self-stresses on rectangular coordinate paper.

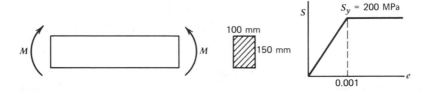

4. The notched member shown below had a strain gage reading of 0.002 at the notch root when a load of 40 kN was applied. Yielding for this material occurs at a strain of 0.004. The load is increased such that the strain gage reads 0.010 when the load is 80 kN. Determine K_t. Also determine K_ϵ and compare it with values calculated using the linear rule and the Neuber rule. $E = 200,000$ MPa, $\nu = 0.3$.

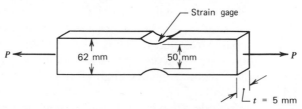

5. Rework the example problem in Section 7.7 for K_t equal to (a) 2.5, (b) 1.5, and (c) 1.0.

CHAPTER **8**

LIFE ESTIMATES FOR CONSTANT AMPLITUDE LOADING

Innumerable constant amplitude tests with axial loading, bending, or torsion have provided a wealth of data on which life estimates can be based. They have taught us that there is no single method that can be applied to all situations. In very sharply notched parts, life estimates are based entirely on crack growth laws. In small smooth laboratory specimens most of the life is spent in the early stages of crack formation so that crack growth may be neglected. These two situations can be handled with confidence. The intermediate case is more difficult for two reasons: (1) the growth of cracks in a region with high stress gradients is hard to compute and (2) it is difficult to tell *a priori* at what crack length one should switch from the shear-dependent first stage to the tension-dependent last stage. A method for dealing with this situation is also presented. None of the methods are "exact." Fatigue lives may vary as much as 10:1 or more for the same loading and material. At best the calculated estimates will only result in a prediction for median life. To obtain a suitably low probability of failure, one must count on a much shorter life.

This chapter shows how to calculate an estimated life to failure for uniaxial constant amplitude loading. For the more frequent cases of variable amplitude loading, which are discussed in Chapter 10, one needs to know damage fractions or cycle ratios, which are the ratio of cycles at given maximum and minimum stresses to the life to failure expected from the same loading. For multiaxial stressing, which is discussed in Chapter 9, one can proceed by "equivalent" uniaxial stresses, which would give the same life. The methods explained in this chapter thus serve the ends of Chapters 9 and 10.

Although constant amplitude uniaxial loading is relatively rare outside the laboratory, a good qualitative understanding of the estimating methods is of value in deciding on design changes for longer fatigue life or for changes in material or dimensions.

We discuss four approaches to fatigue life estimation for constant amplitude loading.

1. A simplified approach that involves one estimate for long life, one for short life, and interpolation between these two.
2. A simplified approach to low cycle fatigue.
3. A two stage approach in which the life to the appearance of visible cracks is estimated from local strain considerations and the life from appearance of small cracks to final fracture is estimated from crack growth considerations.
4. A preexisting crack, and therefore only crack growth is considered.

Before proceeding, a *warning* is in order: The estimates cannot be more accurate than the data and the equations. The needed data are often missing or subject to considerable scatter. Only rarely do we have enough data to estimate a low probability of failure. The equations are based on assumptions and empirical relations, not on well established laws of nature. Estimates obtained by methods explained below are quite sufficient for design of prototypes, but they are *not safe* estimates, at best they estimate only median life.

It would be useless to try for high precision in estimating median fatigue life. As we are ignorant of the tails of the distribution curves we must include a margin of safety or a safety factor in the design calculations. When the safety factor is chosen with a precision of about 20 percent (2 or 2.5; 4 or 5, etc.) there is no point in knowing fatigue strengths corresponding to median life within 5 percent. Estimates with higher precision can be worthwhile only if codes or customers prescribe definite loads and safety factors based on median fatigue life.

8.1 A SIMPLIFIED APPROACH

This approach is based on the use of available or estimated Haigh diagrams (constant life diagrams). These diagrams show the combinations of nominal alternating stress, net regional mean stress,* and notch factor that correspond to the fatigue limit or to a life of 5 million cycles. From such diagrams one obtains one point of an S–N curve. A second point for the S–N curve is obtained from knowledge or estimate of the stress corresponding to a very short life, usually either 1 or 1000 cycles. These points are joined by a straight line on log S–log N coordinates. When test data are available one will of course use them. When such data are not available one must rely on

---

* Net regional mean stress equals nominal mean stress plus or minus self-stress near the surface.

estimates. The next section discusses how to estimate a Haigh diagram. Then the estimates for a very short fatigue life are considered.

8.1.1 Estimating Haigh Diagrams for Long Life

The following data must be either available or estimated by methods that are explained later:

The monotonic yield strength, S_y.

The cyclic yield strength, S_y'.

The unnotched fully reversed fatigue limit, S_f or the fatigue strength at 10^7 reversals.

The true fracture stress, σ_f.

The critical alternating tensile stress, S_{cat}.

The fully reversed long life fatigue notch factor, K_b.

As an example, the Haigh diagram for 4340 alloy steel, heat treated to 409 BHN (Brinell hardness number) is constructed. Estimates based on rules of thumb are compared with fatigue data from the literature.

Monotonic yield strength, S_y. Handbooks such as *Kent's Mechanical Engineers Handbook* [1] show yield strength as a function of tensile strength for medium carbon steels. The tensile strength can be estimated from the hardness

$$S_u \text{ (MPa)} = 3.5 \text{ BHN} = 1430 \text{ MPa}$$

$$S_u \text{ (ksi)} = 0.5 \text{ BHN} = 205 \text{ ksi}$$

For this tensile strength page 2–18 of *Kent's Mechanical Engineers Handbook* shows

$$S_y = 195 \text{ ksi} = 1345 \text{ MPa}$$

The values of 1430 and 1345 MPa can be compared with the data in SAE J 1099 [2], which are 1469 and 1372 MPa for S_u and S_y, respectively. These latter values are also quoted in Table A.2, but without the hardness number.

The cyclic yield strength. S_y' for this material is estimated to be less than the monotonic yield strength, according to the information from Manson on page 27 and in Fig. 3.12. An estimate might be

$$S_y' \simeq 0.8 \, S_y = 1075 \text{ MPa}$$

Table A.2 shows

$$S_y' = 827 \text{ MPa}$$

The difference is substantial. It would be important for considerations of settling of springs, but for our present purpose it is not disturbing.

The next value to be determined is S_f, *the long-life unnotched fully reversed fatigue limit*. This is a key value, more important than S_y or σ_f. If we had no other information we would estimate S_f as a fraction of S_u. This fraction has all too often been taken from Eqs. 5.8a and 5.8b as 0.5 for $S_u \le 1400$ MPa (200 ksi). The fraction, however, can be as low as 0.2 or as high as 0.65. Table A.2 gives these fractions for several types of steel. If the type in question were not listed we might consider the 4142 and 9262 quenched and tempered steel listed as similar. The fractions for these materials are: 0.36, 0.31, and 0.38. Their average is 0.35, which leads to the estimate

$$S_f = 0.35 \times 1430 = 500 \text{ MPa}$$

For comparison Table A.2 shows

$$S_f = 467 \text{ MPa}$$

Table A.1 shows $S_f = 470$ MPa for a slightly harder steel and Grover [3] shows, for $S_u = 208$ ksi

$$S_f = 70 \text{ ksi} = 480 \text{ MPa}$$

Our estimate for S_f seems reasonable. Note that it is less than the 0.5 fraction recommended so frequently.

The next item to be estimated is the *true fracture stress*, σ_f. It provides a measure of the mean stress influence on smooth specimens. High accuracy is not required for our purpose. Table A.2 gives the true fracture stress

$$\sigma_f = 1560 \text{ MPa}$$

It might have been estimated from the rule of thumb [4], $\sigma_f = S_u + 50$ ksi $= S_u + 350$ MPa to be $1430 + 350 = 1780$ MPa. If no information at all is available one can assume, for the purpose of constructing a Haigh diagram only, $\sigma_f \approx 3S_f$, which would give $\sigma_f = 1500$ MPa. The factor of 3 was chosen by noting that the slope of the line connecting S_f and σ_f in a Haigh diagram is roughly $-\frac{1}{3}$. As is mentioned earlier the value of σ_f need not be known with great precision for the present purpose.

The critical alternating tensile stress, S_{cat}. This term is the stress below which cracks will not propagate. It is closely related to the threshold stress intensity range. It was estimated by Fuchs [5] as 70 MPa (10 ksi) for hard steel, 30 MPa (4 ksi) for mild steel, and 20 MPa (3 ksi) for high strength aluminum. However, if any margin for safety us used, S_{cat} will be practically zero. For the purposes of this chapter it can be estimated as 50 MPa (7 ksi).

The fatigue notch factor, K_b. This term is discussed earlier in Chapter 6. For the example Haigh diagram, a notch with $K_b = 2$ is assumed.

We are now ready to construct the Haigh diagram, Fig. 8.1. The abscissa represents mean stress (net regional mean stress). The ordinate represents alternating stress. The points located on these axes are from the estimates (and from literature):

$$S_y = 1345 \ (1372) \ \text{MPa} \qquad S_y' = \pm1075 \ (827) \ \text{MPa}$$

$$S_f = \pm500 \ (467) \ \text{MPa} \qquad \sigma_f = 1780 \ (1560) \ \text{MPa}$$

These four points define the behavior of smooth laboratory specimens. Lines connecting the estimated points are drawn in Fig. 8.1 from $-S_y$ to S_y' to $+S_y$ and from S_f to σ_f, which intersects the first set of lines at A and B. Any combination of mean and alternating stresses outside the triangle from $-S_y$ to S_y' to $+S_y$ corresponds to gross yielding. Any combination above the line A–B will produce a median fatigue life less than 5 million cycles.

For a notched part three additional lines are drawn. One is parallel to AB, with all values divided by K_b. It defines the development of cracks by local high stress. The other two lines correspond to S_{cat}. They are shown as 50 MPa. With tensile mean stress of more than 50 MPa, an alternating stress of 50 MPa is all tensile alternating [$S_{cat} = (S_{max} - S_{min})/2$ for $S_{min} > 0$]. With compressive mean stress, S_m, the alternating stress must be $S_m + (2 \times 50)$ to produce 50 MPa alternating tensile stress. The line corresponding to S_{cat} is horizontal in most of the area of tensile mean stress; the line is inclined at 45 degrees in the area of compressive mean stress. Small cracks will not propagate as long as the combinations of mean and alternating stress are below these two lines.

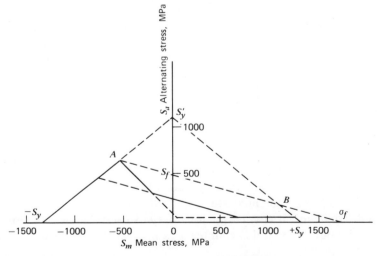

FIGURE 8.1 Estimated Haigh diagram for a part made of 4340 steel, heat treated to 409 BHN, with a fillet giving a fatigue notch factor $K_b = 2$.

For a part with a fillet giving a notch factor $K_b = 2$, the estimated Haigh diagram is shown in solid lines. The noteworthy feature is that maximum alternating stress can be tolerated when the mean stress is compressive and equal to about half the yield strength. Under these conditions the fatigue notch factor is unity and the notch has no effect. The presence of tensile mean stress reduces the amount of alternating stress that can be tolerated. In the region between about -200 and $+700$ MPa mean stress, the fatigue line for the fillet is parallel to the unnotched line, but it is lowered by a factor $K_b = 2$. In terms of alternating stress the notch factor increases to values greater than K_b, and possibly greater than K_t.

With a diagram like Fig. 8.1 we can estimate the long-life fatigue strength of parts with notches for any combination of mean stress and alternating stress. Another method for estimating Haigh diagrams is derived from the fatigue parameter proposed by Smith et al. [6] from Eq. 5.15, which is

$$(\sigma_{max}\epsilon_a E)^{1/2} = \text{constant}$$

With the Neuber relation $\sigma_a\epsilon_a = K^2 S_a e_a$ extended to include $\sigma_m = KS_m$ this becomes

$$K(S_{max}e_a E)^{1/2} = \text{constant}$$

where K can represent K_t, K_f, or K_b, and as long as the nominal alternating stress does not exceed the yield strength

$$K(S_{max}S_a)^{1/2} = \text{constant}$$

and with

$$S_{max} = S_a + S_m$$

and

S_f = fully reversed smooth specimen fatigue strength

$$(S_a + S_m)S_a = \frac{S_f^2}{K^2} \tag{8.1}$$

This relation was used to plot Fig. 8.2 (the continuous curves) for the same material and notch factor as Fig. 8.1. According to this equation, S_a asymptotically approaches zero for large tensile values of S_m and asymptotically approaches S_m for large compressive values of S_m. The Haigh diagram Fig. 8.1, made up straight lines, is superimposed on Fig. 8.2 for comparison. For the notched part the difference between the two approaches is less than our uncertainty. The Haigh diagram with the three criteria is more flexible; the Smith-Topper approach is simpler to program on a computer. Thus we have two methods for estimating a long-life point of the S–N curves for any notch factor and any combination of mean stress and alternating stress.

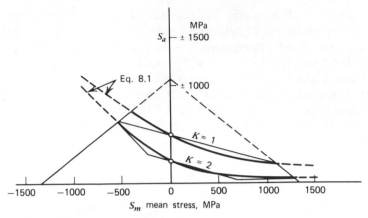

FIGURE 8.2   Haigh diagrams according to $K^2 (S_{max} S_a) = S_f^2$ compared to three criteria Haigh diagrams for $K = 1$ and $K = 2$.

Critical readers may have noted that the yield condition $S_m + S_a = S_y$ based on nominal stresses corresponds to $\sigma_m + \sigma_a > S_y$ for the local stresses at the notch root. This is true, but if the mean stress, usually self-stress, extends to a reasonable depth it will arrest cracks. Because of self-stress concentration in notches and because of the stress gradient in notches the nominal stress is an excellent approximation of the relevant regional stress as shown by Fuchs [7]. The high strength obtainable at point A of Fig. 8.1 also confirms the validity of this approach. The long-life estimated by these simplified methods thus includes both the prevention of crack formation (by limiting the alternating shear stress) and the prevention of crack growth (by limiting the alternating tensile stress).

A point on the S-N curve can thus be estimated for long fatigue life. Estimation for other points, for very short life, follows. Before proceeding to that, however, we should consider the uncertainties that have shown up. Data from different sources may not agree and methods proposed by different investigators may not agree. The uncertainties are caused by the inherent scatter in test data, by the great influence of small details on fatigue life, and by our lack of quantitative data about the basic mechanisms that operate in developing a small fatigue crack. These uncertainties should not deter us from making estimates, but they should cause us to allow sufficient margins for safety and to avoid overelaborate computations that might obscure the core of the matter behind the complexity of the equations and the number of digits printed out by the computer.

8.1.2. Estimating Stress for a Very Short Life

Static fracture is one point that can be used for an estimate of a short life at a high stress. It may be considered to correspond to 1 cycle, 1 reversal, or $\frac{1}{4}$ cycle of fully reversed loading. For our purpose the difference between $\frac{1}{4}$ and 1 cycle can be disregarded. The best estimate is based on a monotonic fracture test. A conservative estimate for ductile metals equates the nominal fracture stress S in the part to S_u, the ultimate tensile strength. Notched parts may show different nominal stresses at fracture than smooth tensile test bars, and extrapolation of fatigue test data of smooth ductile specimens to life at $\frac{1}{2}$ cycle usually shows a higher strength than S_u [2]. The extrapolated strength is called σ_f', the fatigue strength coefficient. It can be estimated to be equal to the true fracture strength, but is not necessarily equal to it as seen in Table A.2. More accurate predictions can be obtained by assuming that smooth and notched parts of ductile materials have equal nominal stresses at 1000 reversals or 500 cycles [8, 9].

Another estimate for a point on the S–N curve at a short life is shown in Fig. 5.18. A strain amplitude of 1 percent corresponds to a life of about 1000 cycles. The stress corresponding to this strain amplitude must be obtained from a cyclic stress–strain curve. For 4340 steel at 209 BHN, for instance, we obtain the stress for 1 percent strain amplitude by a few iterations of the equation

$$\epsilon = \epsilon_e + \epsilon_p = \frac{\sigma}{E} + \left(\frac{\sigma}{K'}\right)^{1/n'} \simeq \frac{\sigma}{E} + 0.002\left(\frac{\sigma}{S_y'}\right)^{1/n'} \tag{8.2}$$

as follows, with the data from Table A.2

σ, MPa	1000	900	950	960	955
ϵ_e, percent	0.5	0.45	0.475	0.48	0.477
ϵ_p, percent	0.71	0.35	0.504	0.54	0.522
ϵ, percent	1.21	0.8	0.979	1.02	0.999

The stress for 1000 cycles can also be obtained from the data in Table A.2 by using the equation

$$\sigma = \sigma_f' \, (2N)^b \tag{8.3}$$

derived from Eq. 5.10 with $\sigma = E \, \Delta\epsilon_e/2$ thus

$$\sigma]_{N=1000} = 2000(2000)^{-0.091} = 1001 \text{ MPa}$$

The difference between the two values, 1001 and 955, is typical of the uncertainties with which the user of fatigue data is faced. It is likely that even a series of fatigue tests would not reduce these uncertainties below a standard deviation of about 5 percent of the stress.

8.1.3 Interpolation

After a stress for long life, of the order of millions of cycles, and a stress for 1 or 1000 cycles have been estimated, one can interpolate by assuming a straight line between these points on an S–N diagram on logarithmic scales. This corresponds to the power law given by Eq. 5.11.

$$(2N)^b = \frac{\sigma_a}{\sigma_f'} \tag{8.4a}$$

and therefore the fatigue lives at two stress amplitudes are in the ratio

$$\frac{N_2}{N_1} = \left(\frac{\sigma_1}{\sigma_2}\right)^{-1/b} \qquad (8.4b)$$

This can be used to calculate the stress corresponding to a desired life N_2 if the life N_1 for stress σ_1 is known. The use of $2N$ instead of N has become common in the literature.

The exponent $-b$ is a small negative number, of the order of -0.1 for unnotched specimens. The exponent $-1/b$ is about 10. Equations 8.4 say that for unnotched specimens the life is approximately inversely proportional to the tenth power of the stress. A 10 percent increase in stress will then reduce the life by approximately one-half; a 10 percent decrease in stress will approximately double the life of most unnotched specimens. For notched parts the slope of the S–N curve on logarithmic scales is steeper.

8.1.4 Examples of Life Estimation

A notched part for which fatigue data have been published is shown in Fig. 8.3. It is made from quenched and tempered hot-rolled steel RQC-100. The elastic stress concentration factor K_t is 3. The nominal stress $P/A + Mc/I$ is $11.32P$ (MPa), where P is the load in kilonewtons, or $7.3P$ (ksi), where

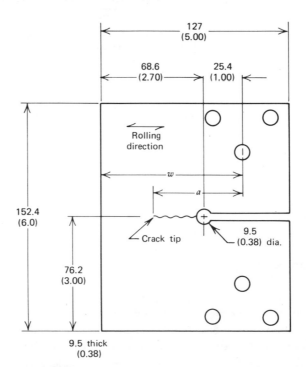

FIGURE 8.3 Test specimen design, mm (in.).

P is the load in kips. Relevant data, from two sources, for RQC-100 in the long transverse direction are given below.

Source	S_u, MPa	σ_f, MPa	σ_f', MPa	b
J1099 [2]	938	1069	1241	−0.07
Tucker and Bussa [10]			1165	−0.075

The reversed fatigue strength at 5 million cycles can be calculated from Eq. 8.3 as

$$1241(10^7)^{-0.07} = 402 \text{ MPa}$$

or

$$1165(10^7)^{-0.075} = 348 \text{ MPa}$$

We note that there is a difference of 15 percent between these values and we use 375 MPa as the long-life point. For the stress at one reversal we use 1200 MPa, approximately the mean between the two published values of σ_f'. These two points establish the $S-N$ curve for smooth specimens as a straight line on log–log coordinates. For the notched part with $K_t = 3$ and a notch radius of 4.75 mm (3/16 in.) we calculate K_b from Neuber's formula (Eq. 6.5) with $a = 0.09$ mm

$$K_b = 1 + \frac{2}{1 + \sqrt{a/r}} = 2.76$$

The long-life reversed fatigue strength of the notched part then is estimated to be 375/2.76 = 136 MPa.

The estimated points, 1200 MPa at $\frac{1}{2}$ cycle, 375 and 136 MPa at 5 million cycles, are plotted on log–log coordinates as alternating stresses versus reversals (1 cycle = 2 reversals) in Fig. 8.4. A straight line from point A, 1200 MPa at 1 reversal, to point B, 375 MPa at 10 million reversals, is the $S-N$ curve for smooth specimens. The $S-N$ curve for the notched part is drawn in two alternative ways as straight lines. One assumes equal strength of notched and smooth parts at 1 reversal; the other assumes smooth and notched parts have equal strength at 1000 reversals. The former assumption would be more correct for brittle materials, the latter for ductile materials. The first of these lines goes from point A to point C at 10 million reversals and 375/2.76 = 136 MPa; the second goes from point C to point D, at 1000 reversals on the line $A-B$.

Test data for the notched part are also shown in Fig. 8.4. They indicate that we either used too high a value for S_f or too low a value for the strength reduction caused by the notch and the surface conditions or both. Crossing of the line for notched parts and that for smooth specimens at 1000 reversals appears to be justified in this instance as two test points are near the crossing. The discrepancies between estimated and experimental values and among data from different sources are typical of fatigue analysis at our current level of knowledge.

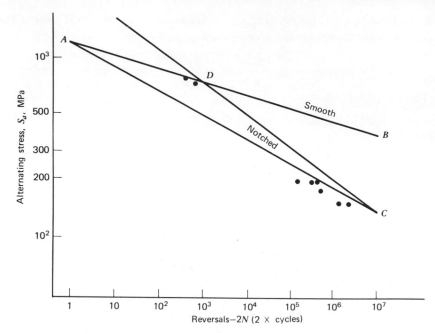

FIGURE 8.4   Estimated and measured fatigue life.

This example was chosen because the test results and materials data are well documented in the literature. It does not show the need for and the use of the Haigh diagram, which is needed when the mean stress is different from zero. Let us assume that the drilled hole (9.5 mm or 0.375 in. diameter) has been treated by expansion or shot-peening to have a self-stress of 700 MPa compression near its surface and apply the estimating method to this case. $S_y = 900$ MPa, $S_y' = 600$ MPa, $\sigma_f = 1069$ MPa from Table A.2; $S_f = 375$ MPa, $K_b = 2.76$ as estimated above, and $S_{cat} = 30$ MPa is assumed. The diagram is shown in Fig. 8.5.

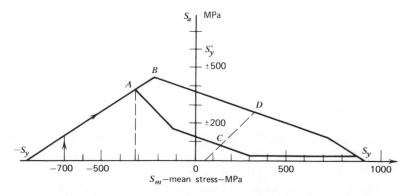

FIGURE 8.5   Haigh diagram for the example part.

The maximum alternating stress that the part is expected to withstand for 5 million cycles is indicated as point A (385 MPa). Note that application of this alternating stress (385 MPa) will reduce the mean stress to 320 MPa because the compressive peak exceeds the yield strength as long as the mean stress is greater than 320 MPa. We indicate this by the arrows in Fig. 8.5. The alternating bending stress for 5 million cycles of a smooth part with optimum compressive self stress near the surface would be point B (450 MPa). With these points and $\sigma_f' = 1200$ MPa (as above) we can construct an estimated S–N line as shown in Fig. 8.6.

As a third example we assume that the same part will be subjected to load cycles from a constant minimum nominal stress of 50 MPa to a maximum nominal stress that will be adjusted to give the desired fatigue life. We want to construct the S–N line.

On the Haigh diagram, Fig. 8.5, a line of constant minimum stress is straight and inclined at 45 degrees upwards to the right. Starting at 50 MPa mean and zero alternating stress we intersect the constant-life line for the notched part at C (90 MPa) and that for the smooth part at D (260 MPa). We construct the S–N line for the smooth part from 260 MPa alternating stress at 10^7 reversals to 1200 MPa at 1 reversal. For the notched part we draw the line from 90 MPa to 1200 MPa at one reversal. These lines are also shown in Fig. 8.6.

The nominal alternating stress at 1 reversal is 1200 MPa for these examples because the mean stress disappears in these parts when the deformation

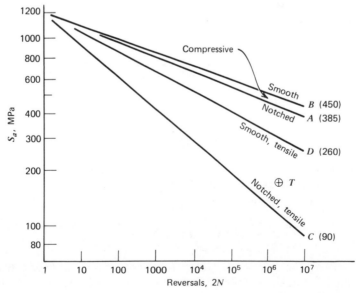

FIGURE 8.6 S–N lines for smooth and notched parts, with tensile and compressive mean stress.

is mainly plastic. Initial self-stress can also be reduced to practically zero
by stretching parts well beyond the onset of yielding. This technique is used
in the fabrication of aluminum plates for sculptured aircraft wings to avoid
distortion by machining. When the outer fibers of solid sections loaded in
bending or torsion are strained beyond yielding, initial mean and self
stresses can also be reduced to practically zero in the surface region. This
is not true for mean stress for axial tests where the whole section must
support the mean load. A single test point for this loading of the notched
part is available in the literature [10]. It is shown as point T in Fig. 8.6. The
agreement is about as usual in fatigue life predictions unless they are fitted
to specific test data. When specific test data can be used to interpolate a
prediction they are, of course, the best available information.

8.2   STRAIN BASED APPROACH FOR LOW CYCLE FATIGUE

Gas turbines operate at fairly steady stresses, but when they are started or
stopped they are subjected to a very high stress range. At the time of
intensive development of jet engines much research on low cycle fatigue
was done. The best approach to this subject is by way of strains rather than
stresses. The strains will be well above the yield strain and the stresses will
be more difficult to measure or to estimate than the strains.

The simplest estimate for short lives is mentioned earlier in Section 5.4.
At a strain amplitude of 1 percent the fatigue life will be about 1000 cycles
as shown in Fig. 5.18. Another simple approximate relation for short lives
can be expressed by the equation [11]

$$2N = \left(\frac{\Delta \epsilon_p}{2\epsilon_f'}\right)^{1/c} \approx \left(\frac{\Delta \epsilon}{2\epsilon_f}\right)^{-2} \tag{8.5}$$

where ϵ_f is the logarithmic strain at fracture in a monotonic test. It is

$$\epsilon_f = \ln \left(\frac{1}{1 - RA}\right) \tag{8.6}$$

where RA is the reduction of area (expressed as a fraction smaller than
one). The exponent 2 and the monotonic value ϵ_f are approximations to the
values $1/c$ and ϵ_f' in the equation

$$\frac{\Delta \epsilon}{2} = \frac{\sigma_f'}{E} (2N)^b + \epsilon_f' (2N)^c = \frac{\Delta \epsilon_e}{2} + \frac{\Delta \epsilon_p}{2} \tag{8.7}$$

which describes the relation between total strain amplitude and life to the
appearance of cracks, 0.25 to 5 mm long or deep. For short lives we
approximate life to fracture by life to the appearance of such cracks and
neglect the elastic strain range $\Delta \sigma / E$ in comparison to the total strain range

$\Delta\epsilon$. Thus Eq. 8.7 for short lives becomes

$$2N = \left(\frac{\Delta\epsilon}{2\epsilon_f'}\right)^{1/c} \qquad \text{for} \quad N < 2000 \text{ cycles} \qquad (8.8)$$

Values for ϵ_f' and for c for more than 50 types of steel and for six aluminum alloys are given in the *SAE Handbook* [2]. Some of these data are given in Table A.2. Values for c range from -0.39 to -0.84, with 50 out of 60 quoted values between -0.5 and -0.7. The fatigue ductility coefficient ϵ_f' is less than half the monotonic ductility ϵ_f for a third of the listed materials. It is greater than ϵ_f for only a few. Note that life is inversely proportional to the fifth to tenth power of stress range, as shown in Fig. 8.6, but inversely proportional to approximately the second power of plastic strain range.

8.3   THE TWO STAGE APPROACH

With change in emphasis in aircraft design from a safe-life philosophy to a damage tolerance philosophy, the prediction of fatigue crack growth has become more important. Damage tolerance requires that the interval of crack growth between easily detectable size and a critical size must be longer than the interval between inspections. The life of a structure is then considered in two stages:

Stage 1:   To the existence of a crack that is not likely to be overlooked in inspection, "crack initiation."

Stage 2:   From the existing crack to failure. Sometimes the specifications require evidence of safety with given cracks, for instance cracks of 2 mm length. Stage 1 then becomes negligible in comparison to stage 2.

Predictions for either stage 1 or stage 2 require far more data than the simplified life predictions discussed earlier in this chapter. Prediction of stage 1 requires the two coefficients σ_f' and ϵ_f' and the exponents b and c, as well as the notch factor K_b and perhaps the cyclic stress–strain curve. Predictions for stage 2 require the stress intensity factor K for the given loading geometry and crack size, knowledge of the crack growth curve da/dN versus ΔK, and knowledge of the fracture toughness. The uncertainties that we note in connection with the simplified methods earlier in this chapter remain or become worse as more data are required to perform the calculations that lead to the desired estimates.

8.3.1   The Local Strain Approach for Early Stage

Strain based approaches to fatigue problems are widely used at present for several reasons. Strain can be measured and has been shown to be an

excellent parameter for correlating with low cycle fatigue. Strain based
approaches unify the treatment of low cycle and high cycle fatigue. When
one wants to predict the number of cycles to the appearance of cracks of
about 0.25 to 5 mm (0.01 to 0.2 in.) long, one generally uses a strain based
approach. Curves of strain versus numbers of reversals and stress–strain
curves are the basic tools of this approach. The strain–life curves are
developed on logarithmic scales from Eq. 8.7, which is repeated here:

$$\frac{\Delta\epsilon}{2} = \tfrac{1}{2}\,(\Delta\epsilon_e + \Delta\epsilon_p) = \frac{\sigma_f'}{E}\,(2N)^b + \epsilon_f'\,(2N)^c \qquad (8.7)$$

The equation says that life to the appearance of small cracks varies as the
$(-1/b)$ power of the elastic strain range (about the tenth power) and also as
the $(-1/c)$ power of the plastic strain range (about the second power) and
that the total strain range is the sum of its elastic and plastic parts. To solve
this equation for N requires iterations. Within the accuracy provided by the
equation a graphical solution is entirely adequate and easy. We show below
examples of its use, without mean stress and with mean stress. Equation
8.7 assumes zero mean stress. Equation 8.9 allows for mean stress

$$\frac{\Delta\epsilon}{2} = \frac{\sigma_f' - \sigma_m}{E}\,(2N)^b + \epsilon_f'\,(2N)^c \qquad (8.9)$$

where σ_m is the mean stress, positive if tensile, negative if compressive.

Figure 8.7 applies the local strain approach to the (RQC-100 steel) example that is used earlier in this chapter. The values quoted from Table A.2

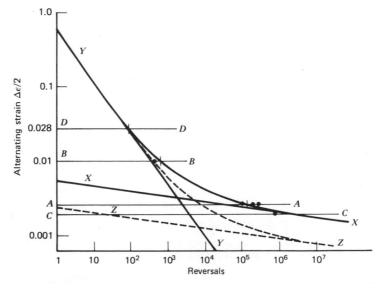

FIGURE 8.7 Strain amplitude versus life to appearance of cracks for the part
shown in Fig. 8.3, made of RQC-100 steel.

are

$$\sigma_f' = 1240 \text{ MPa} \qquad \epsilon_f' = 0.66 \qquad E = 207000 \text{ MPa}$$

$$b = -0.07 \qquad c = -0.69$$

The curve for $\Delta\epsilon/2$ versus $2N$ can now be drawn as the upper solid curve in Fig. 8.7. To locate test data on this chart one converts the given alternating loads to alternating strains at the notch root, according to

$$\frac{\Delta\epsilon}{2} = 1.509 \times 10^{-4} \times \left(\frac{\Delta P}{2}\right)$$

where ΔP is the load range in kilonewtons. This conversion is based on $K = 2.76$, as calculated in Section 8.1.4, on 11.32 MPa/kN nominal stress, and on $E = 207,000$ MPa. Line X–X represents half the elastic strain range, $\Delta\epsilon_e/2$, versus the number of reversals, with a slope of -0.07. As we used double the scale of $2N$ as scale for $\Delta\epsilon/2$, the actual slope on the drawing is -0.14. Line Y–Y represents half the plastic strain range, $\Delta\epsilon_p/2$, versus the number of reversals, with a slope of -0.69 (drawn as -1.38 in our scales). The solid curved line represents the total strain on logarithmic scales. Where the lines for $\Delta\epsilon_e/2$ and $\Delta\epsilon_p/2$ cross, we get a total strain equal to twice the elastic or twice the plastic strain, represented by a point $\log 2 (= 0.3)$ above the intersection. Half the strain range, calculated from the load range, can now be used to intersect the curve for total strain. The estimated life is read on the abscissa. Line A–A is at half the strain range at the notch root for 17.8 kN alternating load (half the load range), which is 0.274 percent. The estimated life is 125,000 reversals. Line B–B is at half the strain range for a 66.7 kN alternating load, 1.02 percent. The estimated life is 630 reversals. Test points for these load ranges corresponding to the development of cracks 2.5 mm (0.1 in.) long are also shown along these lines. The test point for the load range from 4.4 to 31.2 kN is on the line C–C, corresponding to a 13.4 kN alternating load, $\epsilon_a = 0.21$ percent. Mean load of 17.8 kN is allowed for by drawing line Z–Z parallel to X–X and below it, in the ratio $\sigma_m/\sigma_f' = 0.45$, and drawing a new line for total strain, based on Y–Y and Z–Z. This curve is shown by a dashed line.

The procedure explained above is based on the linear assumption $K_\epsilon = K_b$. The strain range was assumed to be proportional to the load range. Use of the stress–strain curve was not required. If the Neuber rule $K_\epsilon = K_b^2/K_\sigma$ is used we must find the strains by using the stress–strain curve. This is done for an alternating load of 66.7 kN in Fig. 8.8. The cyclic stress–strain curve O–X is drawn from the equation

$$\frac{\Delta\epsilon}{2} = \frac{\Delta\epsilon_e}{2} + \frac{\Delta\epsilon_p}{2} = \frac{\Delta\sigma}{2E} + \left(\frac{\Delta\sigma}{2K'}\right)^{n'} \qquad (8.10)$$

with the values $K' = 1150$ MPa (1434), $n' = 0.10$ (0.14), $E = 203,000$ MPa (207,000) given by Tucker and Bussa [10] (values in parentheses from J 1099

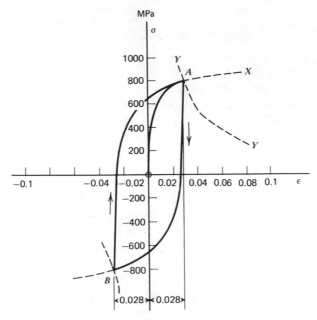

FIGURE 8.8 Hysteresis loop for $\Delta P/2 = 66.7$ kN by Neuber's rule.

for comparison). The nominal stress range $\Delta S/2$ is $11.32 \times 66.7 = 755$ MPa. The nominal strain range is $\Delta e/2 = 755/203{,}000 = 0.0037$. The product $K_b^2(\Delta S/2)(\Delta e/2) = 21.3 = (\Delta\sigma/2)(\Delta\epsilon/2)$. Pairs of values of $\Delta\sigma/2$ and $\Delta\epsilon/2$ that satisfy this equation are

$$\frac{\Delta\epsilon}{2} = 0.01 \quad 0.02 \quad 0.04 \quad 0.08$$

$$\frac{\Delta\sigma}{2} = 2130 \quad 1065 \quad 532 \quad 266$$

These points define the hyperbola, Y-Y. It intersects the stress–strain curve at point A, which satisfies the Neuber rule.

From point A to the next reversal at point B the stress and strain describe half a hysteresis loop. It is commonly assumed that the hysteresis loop is bounded by curves that are geometrically similar to the stress–strain curve but twice as large. Such a curve is drawn. Point B, the next reversal, is located at the intersection of this boundary curve with a new hyperbola, which has its origin at point A and satisfies the equation

$$\Delta\epsilon \, \Delta\sigma = K^2 \, \Delta e \, \Delta S = \text{constant}$$

ΔS is 11.32×133.4 MPa; we now must have

$$\Delta\epsilon \times \Delta\sigma = 85.2 = 4 \times 21.3$$

The curves in Fig. 8.8 were drawn by hand, by laying suitable hyperbolas and loop curves under a transparent sheet and shifting their origins first to A and then to B. Computer programs are available to do this entire procedure [12].

We now can read $\Delta\epsilon/2 = 0.028$ from this figure and enter it on Fig. 8.7 on line D–D. The estimated life is now 90 reversals, less than it was with the linear assumption. The test life, 388 reversals, is between the two estimates. If the Neuber procedure had been applied to the load range of 17.8 kN, the difference between the Neuber assumption and the linear assumption would have been very small because there is very little plastic strain involved at the notch root at that load range. We may compare the estimates with test values as follows:

Alternating Load, kN	Life to Cracking		
	By Test	By Linear Rule	By Neuber's Rule
66.7	388	630	90
17.8	201,000 (av. of 3)	125,000	125,000

As is mentioned earlier, Neuber's rule gives more accurate results when plastic flow is not restrained by surrounding elastic material, as in plane stress. The linear rule gives more accurate results when plastic flow is strongly restrained, as in plane strain. The part used for this example, a plate 9.5 mm (0.375 in.) thick with a notch radius of 4.75 mm (0.188 in.) behaved in an intermediate manner, which has been verified by strain measurement [13].

8.3.2 The Fracture Mechanics Approach for Crack Propagation

The LEFM approach to fatigue crack growth life prediction is given earlier in Sections 4.1 and 5.5. To predict the constant load amplitude crack growth life in the previous example, several pieces of information must be known as follows:

1. The stress intensity factor, K_I.
2. The fracture toughness, K_c.
3. The applicable fatigue crack growth rate expression.
4. The initial crack size, a_i.
5. The final or critical crack size, $a_f (a_c)$.

Several expressions for K_I for the SAE keyhole component have been developed using experimental methods or finite elements [14–16]. The solutions compare very favorably and we use here the finite element solution obtained by Socie [16] and converted to SI units as follows.

$$K_I = P(0.247)\left[-54.98 + 417.28\left(\frac{a}{w}\right) - 939.76\left(\frac{a}{w}\right)^2 + 739.31\left(\frac{a}{w}\right)^3\right]$$

(8.11)

where P is in kilonewtons and K_I is in MPa$\sqrt{\text{m}}$. This expression is applicable for a width B equal to 9.5 mm and a/w between about 0.32 and 0.75, which corresponds to effective crack lengths a as measured from the load line of 30 to 70 mm.

The fracture toughness, K_c, for RQC-100 for this thickness, along with the Paris equation (Eq. 5.17) for fatigue crack growth behavior with $R \approx 0$, was also obtained by Socie [16] and converted to SI units. These values are:

$$K_c \approx 165 \text{ MPa}\sqrt{\text{m}}$$

(8.12)

$$\frac{da}{dN} \text{ (m/cycle)} = 2.80 \times 10^{-12} \, (\Delta K \text{ MPa}\sqrt{\text{m}})^{3.25}$$

(8.13)

Dowling and Walker [17] showed that region II and III fatigue crack growth rate behavior for RQC-100 with $R = 0$ was essentially the same as for $R = -1$ when ΔK is the stress intensity factor range as defined by Eq. 5.16. Thus Eq. 8.13 should provide reasonable results for $R = -1$, which exists in this example.

The strain–life prediction model in the preceding section is for actual crack lengths that vary from 0.25 to 5 mm. The LEFM model must then assume initial crack lengths compatible with this range. A reasonable assumption may be to let the initial crack extension from the notch, Δa_i, equal 2.5 mm, which is consistent with the SAE experimental fatigue program and does satisfy the above criteria. This size crack extending from the keyhole notch also allows the use of Eq. 8.11 for K_I. For shorter crack extension, the K_I solution would have to include a more complex analysis involving a small crack growing from a notch in a local plastic strain field [18–21]. Thus we assume an initial effective crack length a_i of 30.2 mm plus 2.5 or 32.7 mm. This gives an initial a/w value of 0.35. The initial K_{max} for this crack with $P_{max} = 66.7$ kN from Eq. 8.11 is 125 MPa$\sqrt{\text{m}}$. For P_{max} equal to 17.8 kN the initial K_{max} is 33 MPa$\sqrt{\text{m}}$. These values are both well above ΔK_{th} levels and thus the Paris equation is applicable from that standpoint. Calculation of the plane stress initial plastic zone size $2r_y$ from Eq. 4.9

$$2r_y = \frac{1}{\pi}\left(\frac{K_{max}}{S_y}\right)^2$$

(8.14)

gives $2r_y = 7.4$ mm for the high load and 0.5 mm for the lower load using $S_y = 830$ MPa. The use of an LEFM model for the high load is conjectural because of the large plastic zone size relative to the specimen thickness and uncracked ligament. Also, the initial K_{max} value for the high load is already in region III of the sigmoidal curve where the Paris equation gives unconservative life predictions. We use the LEFM model and the Paris equation, however, to obtain a first approximation. A J-integral approach could provide better results for the high load [22].

Critical crack lengths at fracture for the two loadings can be approximated by setting K_I from Eq. 8.11 equal to K_c and solving for a/w using an interative or trial and error procedure. For P_{max} equal to 66.7 kN the critical a/w value is 0.46 or $a_c = 43$ mm. For P_{max} equal to 17.8 kN, the critical a/w value is 0.735 or $a_c = 69$ mm.

The fatigue crack growth life estimation, N_f, can be found by integrating the Paris equation

$$\frac{da}{dN} = A(\Delta K)^n = A(K_{max} - 0)^n \qquad (8.15)$$

or

$$N_f = \int_0^{N_f} dN = \int_{a_i}^{a_c} \frac{da}{A(K_{max})^n} = \frac{1}{2.8 \times 10^{-12}(0.247\, P_{max})^n}$$

$$\times \int_{a_i}^{a_c} \frac{da}{\left[-54.98 + 417.28\left(\frac{a}{w}\right) - 939.76\left(\frac{a}{w}\right)^2 + 739.31\left(\frac{a}{w}\right)^3 \right]^n}$$

$$(8.16)$$

Equation 8.16 must be integrated numerically between the limits a_i and a_c. Equation 5.22f cannot be used since K_I is not just a simple $\sqrt{a}$ function. Integration of Eq. 8.16 with the proper values of A, n, P_{max}, a_i, a_c, and w gives N_f equal to 354 cycles for $P_{max} = 66.7$ kN and 39,000 cycles for $P_{max} = 17.8$ kN. The experimental test data had 6 cycles for the high load and 57,000, 61,200, and 93,800 cycles for three tests, or an average of 70,200 at the low load. As is indicated earlier, the high load crack growth estimation using this model was conjectural. However, the life to grow the crack from $\Delta a = 2.5$ mm to fracture for the high load was negligible, which was shown experimentally and with the first approximation model. The lower load crack growth life estimation was about 60 percent of the average experimental life, which can be considered to be within the expectations for these calculations. Life predictions within a factor of 2 are about all one can expect at present.

8.4 CONCLUSION

Which fatigue life prediction method best serves the intended purpose depends on the shape of the part, the material, and the type of loading. For the part shown in Fig. 8.3 and for the fully reversed loading that is used earlier as an example, the simplified methods came as close to the total fatigue life to fracture as can be expected.

This would not be true if the same notch were in a wide thin plate where the crack can grow much longer. When a design is intended for very long fatigue life, the simplified methods of Section 8.1 may serve best for this purpose.

For a very short fatigue life of parts without initial cracks the strain based approach, Section 8.2, serves best. Crack growth can be so rapid that life to the appearance of cracks is the main concern.

Parts designed for intermediate fatigue life present a more difficult problem of choice of approach. An analysis of crack propagation seems indicated in all cases where cracks can be detected by inspection and the part is large enough that there will be a safe interval between first detection of the crack and complete fracture. The analysis will produce an estimate of this safe interval.

For many parts a fail-safe design philosophy may be more profitable than a very detailed analysis. In this connection fail-safe implies warning before catastrophic failure. "Leak before burst" is one phrase of the fail-safe philosophy. "Yield before break" is another. Many parts on which we rely daily are not "fail-safe." Aircraft landing gears and automobile ball joints are examples. They must be designed by applying analysis techniques to data obtained from extensive experience and from realistic testing and service monitoring.

The methods explained in this chapter are intended to serve as guides for conservative long-life design, as guides for estimating the rate of crack growth where cracks can be tolerated for a limited time, and as guides for modifications based on tests or field experience with existing designs.

8.5 DOS AND DON'TS IN DESIGN

1. Do consider fatigue properties and fracture toughness properties of materials.
2. Don't consider average or median materials data as allowable values in design.
3. Do consider two stages of fatigue life: the production of a crack by the total stress range and the propagation of cracks by the tensile stress range.
4. Don't expect an elaborate complex analysis to obviate the need for tests and service monitoring of machines or components.

REFERENCES FOR CHAPTER 8

1. C. Carmichael, Ed., *Kent's Mechanical Engineers' Handbook,* Design and Production Volume, 12th ed., John Wiley and Son, New York 1950, p. 2.
2. "SAE Information Report J1099 on Fatigue Properties," *SAE Handbook,* 1978, p. 4.44.
3. H. Grover, *Fatigue of Aircraft Structures,* NAVAIR 01-1A-13, Navy Department, 1966.
4. J. A. Graham, Ed., *Fatigue Design Handbook,* SAE, 1968 (from a report by J. Morrow), p. 24.
5. H. O. Fuchs, "A Set of Fatigue Failure Criteria," *Trans. ASME, J. Basic Eng.,* Vol. 87, June 1965, p. 333.
6. K. N. Smith, P. Watson, and T. H. Topper, "A Stress-Strain Function for the Fatigue of Metals," *J. Mater.,* Vol. 5, No. 2, Dec. 1970, p. 767.
7. H. O. Fuchs, "Regional Tensile Stress as a Measure of the Fatigue Strength of Notched Parts," *Proc. 1971 Int. Conf. Mech. Behav. Mater., Kyoto, Japan,* 1972, p. 478.
8. D. V. Nelson and H. O. Fuchs, "Predictions of Cumulative Fatigue Damage Using Condensed Load Histories," *Fatigue under Complex Loading,* SAE, 1977, p. 163.
9. V. M. Faires, *Design of Machine Elements,* Macmillan, London 1965, p. 119.
10. L. Tucker and S. Bussa, "The SAE Cumulative Fatigue Damage Test Program," *Fatigue under Complex Loading,* SAE, 1977, p. 1.
11. J. F. Tavernelli and L. F. Coffin, "Experimental Support for Generalized Equation Predicting Low Cycle Fatigue," *Trans. ASME, J. Basic Eng.,* Vol. 84, No. 4, Dec. 1962, p. 533.
12. R. W. Landgraf, F. D. Richards, and N. R. La Pointe, "Fatigue Life Predictions for a Notched Member under Complex Load Histories," *Fatigue under Complex Loading,* SAE, 1977, p. 95.
13. N. E. Dowling, W. R. Brose, and W. K. Wilson, "Notched Member Life Predictions by the Local Strain Approach," *Fatigue under Complex Loading,* SAE, 1977, p. 55.
14. R. C. Rice, Report to SAE Fatigue Design and Evaluation Committee.
15. D. V. Nelson, "Cumulative Fatigue Damage in Metals," Ph.D. Thesis, Stanford University, 1978.
16. D. F. Socie, "Estimating Fatigue Crack Initiation and Propagation Lives in Notched Plates under Variable Load Histories," Ph.D. Thesis, University of Illinois, 1977.
17. N. E. Dowling and H. Walker, "Fatigue Crack Growth Rate Testing of Two Structural Steels," Westinghouse R & D Center, Scientific paper 77-1D3-PALFA-P3, Nov. 1977.
18. O. L. Bowie, "Analysis of an Infinite Plate Containing Radial Cracks Originating at the Boundary of an Internal Circular Hole," *J. Math. Phys.,* Vol. 25, 1956, p. 60.

19.   J. C. Newman, "An Improved Method of Collocation for the Stress Analysis of Cracked Plates with Various Shaped Boundaries," NASA TN D-6376, Aug. 1971.

20.   N. E. Dowling, "Fatigue at Notches and the Local Strain and Fracture Mechanics Approaches," Westinghouse R & D Center, Scientific paper 78-1D3-PALFA-P1, March 1978.

21.   M. H. El Haddad, K. N. Smith, and T. H. Topper, "A Strain Based Intensity Factor Solution for Short Fatigue Cracks Initiating from Notches," *Fracture Mechanics,* ASTM STP 677, 1979, p. 274.

22.   N. E. Dowling and J. A. Begley, "Fatigue Crack Growth During Gross Plasticity and the *J*-Integral," *Mechanics of Crack Growth,* ASTM STP 590, p. 82.

PROBLEMS FOR CHAPTER 8

1. Show that the criteria for the alternating tensile stress, $S_{at} = S_{cat}$ plots on the Haigh diagram as
$$S_a = S_{cat} \qquad \text{for} \quad R \geq 0$$
and
$$S_a + S_m = 2S_{cat} \qquad \text{for} \quad R \leq 0$$
where S_a and S_m are the applied alternating and mean stress, respectively.

2. A stepped circular rod of Man-Ten steel of 50 mm diameter and 40 mm diameter has a root radius of 2 mm at the stepped section. The rod is to be subjected to axial cyclic loading. Using a Haigh diagram, determine the following for an approximate median fatigue life of 10^7 cycles:

   (*a*) What fully reversed alternating force, P_a, can be applied?
   (*b*) What is the maximum value of P_a if proper compressive self-stresses are present at the notch root. What is the magnitude of the compressive self-stress needed for obtaining this maximum alternating stress?
   (*c*) What value of P_a can be applied if the self-stress calculated in *b* is tensile?

3. Same as Problem 2 except the material is as follows:

   (*a*) RQC-100 steel.
   (*b*) 4340 steel with $S_u = 827$ MPa.
   (*c*) 4340 steel with $S_u = 1240$ MPa.
   (*d*) 4340 steel with $S_u = 1470$ MPa.
   (*e*) 4340 steel with $S_u = 1950$ MPa.
   (*f*) 2024-T3 aluminum.
   (*g*) 5456-H3 aluminum.
   (*h*) 7075-T6 aluminum.

4. In problem 2, estimate the values of P_a for parts *a*, *b*, and *c* for median fatigue lives of 5×10^4 cycles.

5. Repeat problem 4 with the materials given in problem $3a$ to $3h$.

6. A flat plate of 4340 steel Q & T with S_u = 1250 MPa is 80 mm wide, 300 mm long, and 10 mm thick and has a central through hole of 20 mm diameter. A constant amplitude repeated axial force varying from 0 to 400 kN in tension consistently causes premature fatigue failures. It is desired to increase the life to 10^7 cycles without changing the minimum hole size and material. Four ways of increasing the fatigue life would be to change (a) thickness, (b) width, (c) hole geometry, and (d) self-stresses at the notch root. Provide four quantitative solutions using suggestions a to d separately such that fatigue failures should not occur in 10^7 cycles.

7. RQC-100 hot-rolled steel sheet was machined to the shape as shown. The member was subjected to a fully reversed alternating force $P_a = \pm$ 220 kN. A strain gage located at the notch root read a stable strain range of $\Delta\epsilon = 0.02$.

   (a) Using low cycle fatigue concepts determine the expected number of cycles to the appearance of a small crack of length $\Delta a \approx 0.25$ mm.

   (b) Determine a reasonable number of cycles to grow this crack from $\Delta a = 0.25$ mm to fracture (note: $a_i = \Delta a + 2.5$ mm).

   (c) Compare a and b and discuss the significance and variability of each calculation.

   (d) Repeat a to c for strain $\Delta\epsilon = 0.005$ due to increasing the notch root radius.

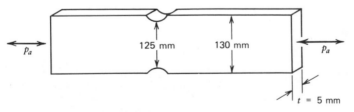

8. Same as problem 7 except the material is Q & T 4142 steel with S_u = 1400 MPa (note: this implies a change in the notch root radius).

9. Same as problem 7 except the material is Q & T 4142 steel with S_u = 1930 MPa (note: this implies a change in the notch root radius).

10. A stepped rotating beam shaft with diameters of 40 and 30 mm and a notch root radius at the step of 5 mm is to be subjected to a constant amplitude bending moment M. The desired life of the component is 50,000 cycles. The shaft is to be made of mild steel. Using the notch strain analysis method, estimate the allowable bending moment using the following:

    (a) Linear rule for K_ϵ.
    (b) Nueber's rule for K_ϵ.

Discuss the difference in results and what value of M you would recommend as the final value. If the above calculations are for cycles to
form a circumferential crack, say, 2.5 mm deep, estimate the remaining
cycles of crack growth to fracture.

11. Same as problem 10 except use 2024-T3 aluminum.
12. Same as problem 10 except use 9262 Q & T steel with S_u = 1000 MPa.
13. Fractographic electron microscopy often reveals indications of fatigue
    crack growth striations that can be used to estimate prevailing stress
    intensity levels. Figure 3.17a shows one such electron fractograph. The
    component was a uniaxially loaded single-edge cracked plate of width
    20 mm, thickness 5 mm, and crack length 6 mm. For R = 0 constant
    amplitude loading, what approximate axial force was applied to the
    plate? Assume the Paris equation constants are A = 10^{-12} and n = 3.5,
    where da/dN is in m/cycle and ΔK is in MPa$\sqrt{m}$.
14. For problem 1 of Chapter 4 use a computer numerical integration pro-
    gram and determine the approximate number of cycles to grow the
    crack to fracture. Some material properties may have to be estimated
    or found elsewhere. The following plate materials are used:

    (a) Mild steel.
    (b) D6aC.
    (c) 2024-T3 aluminum.
    (d) Ti–6Al–4Va.

CHAPTER **9**

MULTIAXIAL
STRESSES AND STRAINS

Multiaxial states of stress are very common. Multiaxial strain is difficult to avoid. The strains are triaxial, for example, in a tensile bar. With longitudinal strain e we have two transverse strains $-ve$, where v is Poisson's ratio, which changes from about 0.3 in the elastic range to 0.5 in the plastic range. In a shaft that transmits torque, we have two principal stresses S, equal in amount but opposite in sign. There is neither stress nor strain in the third principal direction. In a crankshaft we have torsion and bending. On points of its surface we have two principal stresses that vary in magnitude and direction. The frequencies of the bending cycles and torsion cycles are not the same. Most points on the surface have triaxial strain.

The state of stress in notches is often not the same as the state of stress in the main body, and the stress concentration factor changes with the state of stress. A transverse hole in a shaft in torsion produces a stress concentration. On the surface of the hole the state of stress is uniaxial, although it is biaxial in the shaft. At the bottom of a circumferential groove or thread the state of stress is biaxial, although it may be uniaxial in the main body of a round bar or bolt.

Can we somehow manage to apply data from uniaxial tests to multiaxial situations? This is the question with which this chapter is concerned.

To handle this complex subject we break it down into several classifications as follows:

Yielding
Crack initiation
    Under simple multiaxial stresses or strains (see p. 175)
    Under complex stresses or strains (see p. 182)
Crack growth

In these discussions we assume isotropic material, although we know that the fatigue properties of many products are not isotropic, for example, the fatigue strength in the direction of rolling or extruding may be substantially greater than in the transverse direction. We also assume uniformly repeated cycles, unless the opposite is stated.

9.1 YIELDING UNDER MULTIAXIAL STRESSES OR STRAINS

The theory of yielding is covered in books on plasticity and on the mechanical behavior of materials [1, 2]. We review it briefly because it is needed in following sections. Time effects such as creep and viscoelasticity are disregarded, so that yielding depends only on instantaneous increments of stress or strain and on the previous history of the material. The yield criterion can be visualized in a "stress space" in which each of the coordinate axes represents one principal stress. Each point in this space corresponds to a state of stress, defined by the three principal stresses. The yield criterion that we use can be visualized as a circular cylinder in the stress space. For virgin material the axis of the cylinder passes through the origin of the coordinates. It is inclined equal amounts to the three coordinate axes. It represents pure hydrostatic stress. The distance of a point from this axis is a measure of the magnitude of the stress deviator. It has the three components D_1, D_2, and D_3. The cylinder is the locus of points of constant magnitude of the stress deviator $(D_1{}^2 + D_2{}^2 + D_3{}^2)^{1/2}$ = constant. The three components are:

$$D_1 = \sigma_1 - \tfrac{1}{3}(\sigma_1 + \sigma_2 + \sigma_3)$$

$$D_2 = \sigma_2 - \tfrac{1}{3}(\sigma_1 + \sigma_2 + \sigma_3) \tag{9.1}$$

$$D_3 = \sigma_3 - \tfrac{1}{3}(\sigma_1 + \sigma_2 + \sigma_3)$$

For biaxial states of stress ($\sigma_3 = 0$), the yield condition is the intersection of this cylinder with the σ_1-σ_2 plane, which is the well-known yield ellipse with principal axes in the ratio $\sqrt{3}:1$. It is often convenient to convert the multiaxial stress state to an "equivalent" stress, which is the uniaxial stress that is equally distant from (or located on) the yield surface. For virgin material the condition is:

$$\sigma_e = \frac{[(\sigma_1 - \sigma_2)^2 + (\sigma_2 - \sigma_3)^2 + (\sigma_3 - \sigma_1)^2]^{1/2}}{\sqrt{2}} \tag{9.2}$$

or

$$\sigma_e = \frac{[(\sigma_x - \sigma_y)^2 + (\sigma_y - \sigma_z)^2 + (\sigma_z - \sigma_x)^2 + 6(\tau_{xy}^2 + \tau_{yz}^2 + \tau_{zx}^2)]^{1/2}}{\sqrt{2}} \tag{9.3}$$

where σ_1 and so on are the principal stresses and σ_x and so on are stresses in an arbitrary orthogonal coordinate system.

These expressions for the yield equivalent stress σ_e are proportional to a number of quantities, and sometimes these quantites and their names are used in the literature. These quantities are as follows:

The octahedral shear stress $(\sqrt{2/3})\sigma_e$.

The magnitude of the stress deviator $\sqrt{2}\,\sigma_e$.

The distortion energy $\frac{1}{6}(\sigma_e^2/G)$.

The root mean square of the shear stresses on all the planes through a point $= \sqrt{2/15}\,\sigma_e$.

The second invariant of the stress tensor $\frac{1}{3}\sigma_e^2$.

The use of these quantities as yield criteria was proposed independently by von Mises [3], Henky [4], Huber [5], and earlier by Maxwell. The root mean square relation was published by Novozhilov [6].

The yield condition or yield criterion that we describe above is most generally accepted at present. Other yield criteria are discussed by Nadai [1] and Dieter [2].

The expression "virgin material" means material that has not been strain hardened. The effect of strain hardening is described by a "hardening rule." One possible hardening rule is the isotropic rule, which assumes that strain hardening corresponds to an enlargement of the yield surface without change of shape or position in the stress space. Another is the kinematic rule, which assumes that strain hardening shifts the yield surface without changing its size or shape. According to the isotropic rule there is no Bauschinger effect. According to the kinematic rule the sum of compressive and tensile yield stresses is a constant.

To relate stresses and plastic strains we also need a flow rule. We use the following in connection with Garud's method in Section 9.3.1: *The increment of plastic strain caused by an increment of stress is such that the ratios of the principal plastic strain increments are equal to the ratios of the principal values of the stress deviator. Their magnitudes follow from the hardening rule and the uniaxial stress–strain curve.*

Detailed calculations of yielding are needed for the analysis of low cycle fatigue under complex multiaxial loadings such as one sees in pressure vessels and crankshafts.

9.2 CRACK INITIATION IN SIMPLE MULTIAXIAL LOADING

The early stages of fatigue, which are here called crack initiation, depend primarily on the ranges of shear stresses, or on the plastic work. The theories, therefore, have much in common with the theories of yielding, which also depends on a measure of shear stresses. A general theory is discussed in Section 9.3. A much simpler method can be used for those cases in which the principal alternating stresses do not change their directions relative to the stressed part. We call these simple multiaxial stresses. When the directions change we say we have complex multiaxial stresses.

For simple loading two theories fit most observations for long life fatigue and can be extended for application to strain-controlled low cycle fatigue. Sines' method [7] is easier to apply. Langer's method [8] has been incorporated in some codes.

9.2.1   Sines' Method

This method uses the alternating octahedral shear stress. It can be represented by a formula of the type

$$C = S + kH \qquad (9.4)$$

where C = fatigue criterion; equal C corresponds to equal life and higher C corresponds to shorter life
   S = alternating octahedral shear stress
   k = an empirical factor
   H = octahedral normal stress, which is also hydrostatic stress

The method can be expressed by the equation:

$$[(S_{a1} - S_{a2})^2 + (S_{a2} - S_{a3})^2 + (S_{a3} - S_{a1})^2]^{1/2}$$
$$+ \ m(S_{m1} + S_{m2} + S_{m3}) = \frac{\sqrt{2}\, S_N}{K_f} \qquad (9.5)$$

For biaxial stresses, this reduces to:

$$(S_{a1}^2 - S_{a1}S_{a2} + S_{a2}^2)^{1/2} + \frac{m(S_{m1} + S_{m2})}{\sqrt{2}} = \frac{S_N}{K_f} \qquad (9.6)$$

where S_{ai} = alternating component of the principal nominal stresses
   S_{mi} = mean component of the principal nominal stress
   S_N = uniaxial fully reversed fatigue stress that is expected to give the same life on smooth specimens as the multiaxial stress state
   m = coefficient of mean stress influence; it is of the order of 0.5
   K_f = fatigue notch factor; it permits the use of nominal stresses in Eqs. 9.5 and 9.6

If the theoretical stress concentration factors differ too greatly for different principal directions, one must calculate and use the local principal stresses on the left side of these equations and omit K_f on the right side.

Areas of "safe" alternating stress are shown schematically in Fig. 9.1 on a stress plane for biaxial stresses. Lines of constant octahedral shear stress are ellipses in such a stress plane. For any biaxial, in-phase alternating

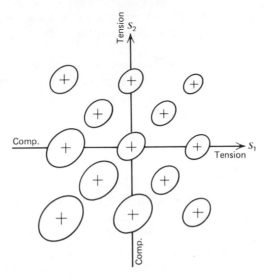

FIGURE 9.1 Areas of "safe" alternating stress for biaxial stresses.

stress that corresponds to the fatigue criterion, successive instantaneous states of stress are represented by points on a diameter of such an ellipse. The center of the ellipse is the mean stress. Ellipses that are located on a line of constant hydrostatic stress (45 degrees downward to the right) are of equal size. Increasing hydrostatic mean stress decreases the size of the ellipse. (Hydrostatic mean stress is equal to octahedral normal mean stress.)

While the field of ellipses is useful for visualization, the use of "equivalent" uniaxial stresses is more convenient and avoids use of the factor m. Equivalent stresses S_{qa} and S_{qm} are those alternating and mean uniaxial stresses that can be expected to give the same life as the given multiaxial stresses. The equivalent alternating stresses according to Sines' theory are:

$$S_{qa} = \frac{[(S_{a1} - S_{a2})^2 + (S_{a2} - S_{a3})^2 + (S_{a3} - S_{a1})^2]^{1/2}}{\sqrt{2}} \qquad (9.7)$$

where S_{a1}, S_{a2}, and S_{a3} are principal alternating stresses.

The equivalent mean stress is the sum of the mean normal stresses in any three mutually perpendicular directions

$$S_{qm} = S_{mx} + S_{my} + S_{mz} \qquad (9.8)$$

After S_{qa} and S_{qm} are calculated the expected fatigue life is found from constant life diagrams or from the formulas for uniaxial fatigue. The formulas and diagrams are used with the magnitude of S_{qa} in place of S_a and the magnitude S_{qm} in place of S_m.

Two examples of the application of Sines' theory may be useful. A closed-end tube, with inside diameter $d = 50$ mm (2 in.) and wall thickness $t = 3$

mm (0.125 in.) is subject to internal pressure (P), which fluctuates from 0 to 30 MPa (4350 psi). How do we compare this stress situation to test data from uniaxial stresses?

Stress analysis shows a longitudinal stress S_L varying from 0 minimum to $Pd/4t = 30 \times 50/12 = 125$ MPa maximum and a circumferential stress S_C from 0 to $Pd/2t = 250$ MPa maximum.

For fatigue analysis we separate the stresses into mean and alternating components:

$$S_{Ca} = 125 \text{ MPa} \qquad S_{Cm} = 125 \text{ MPa}$$

$$S_{La} = 62.5 \text{ MPa} \qquad S_{Lm} = 62.5 \text{ MPa}$$

We then form "equivalent" alternating and mean stresses. They are equivalent because we expect their joint effect to give the same life in uniaxial tests that we expect from the multiaxial situation. For long life the equivalence is calculated by formulas 9.7 and 9.8. (We still may have to check separately, by different formulas, for equivalence with respect to yielding and low cycle fatigue, to early crack formation, and to crack propagation. The methods for those calculations are explained later.)

The equivalent stress for long life formation of short cracks is calculated from Eq. 9.7.

$$
\begin{aligned}
S_{qa} &= \frac{[(S_{Ca} - S_{La})^2 + (S_{La} - 0)^2 + (0 - S_{Ca})^2]^{1/2}}{\sqrt{2}} \\
&= \frac{62.5\,[(2 - 1)^2 + 1^2 + 2^2]^{1/2}}{1.41} \\
&= 62.5 \times \frac{\sqrt{6}}{\sqrt{2}} = 108 \text{ MPa } (15.7 \text{ ksi})
\end{aligned}
$$

The equivalent mean stress from Eq. 9.8 is simply the sum of the mean normal stresses in three mutually perpendicular directions. In our case $S_{qm} = 62.5 + 125 = 187.5$ MPa (27 ksi).

With these two values we can enter a constant life diagram such as Fig. 5.13 or apply a method of predicting life from given uniaxial mean and alternating stresses.

For another example consider a shaft carrying a gear on its end as in Fig. 9.2. Tangential force on the gear is 8000 N (1800 lb). Radial force is neglected. Dimensions are given below.

Pitch diameter: 110 mm (4.33 in.)

Shaft diameter: 20 mm (0.8 in.)

Distance from gear center to bearing: 25 mm (1 in.)

Bending moment: $M = 8000 \times 0.025 = 200$ N·m

Torque: $T = 8000 \times 0.055 = 440$ N·m

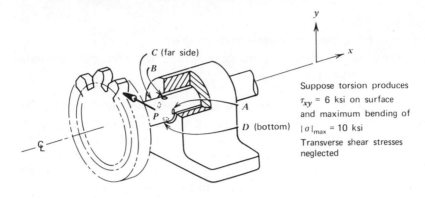

Suppose torsion produces $\tau_{xy} = 6$ ksi on surface and maximum bending of $|\sigma|_{max} = 10$ ksi Transverse shear stresses neglected

Principal stresses at the four surface elements of the shaft shown above

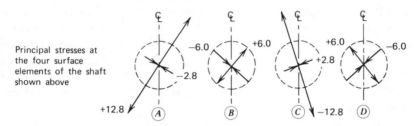

FIGURE 9.2 Multiaxial stressed shaft and gear component.

The stresses at the edge of the bearing are as follows:

Bending stress: $\sigma = 200 \dfrac{32 \times 10^6}{\pi \times 2^3}$ Pa $= 250$ MPa (36 ksi) reversed

Torsion stress: $\tau = 440 \dfrac{16 \times 10^6}{\pi \times 2^3}$ Pa $= 257$ MPa (40 ksi) steady

We calculate S_{qa} and S_{qm} from Eqs. 9.7 and 9.8

$$S_{qa} = \frac{250 \left[(1-0)^2 + 0^2 + (0-1)^2\right]^{1/2}}{\sqrt{2}} = 250 \text{ MPa (36 ksi)}$$

and from the two normal stress components of the steady torsion stress ($\sigma_1 = \tau; \sigma_3 = -\tau$)

$$S_{qm} = 275 - 275 = 0$$

The fatigue life of this shaft is the same as that of a specimen of the same material, with the same notch factor, in uniaxial loading with $S_m = S_{qm} = 0$ and $S_a = S_{qa} = 250$ MPa.

The field of application of Sines' theory is bounded by the following conditions:

No gross yielding.

Stress above the crack propagation threshold.

The alternating stresses can be represented by principal alternating stresses along fixed principal axes.

The data from which Sines' theory was developed included only in-phase alternating stresses, which are represented by straight lines in the stress spaces. It is reasonable to assume that out-of-phase stress cycles, which would be represented by closed curves, could be handled as well. The equivalent alternating stress would be equal to the radius of the inclined cylinder into which the closed curve fits in the stress space.

A few remarks are in order. In our second example the principal axes do not remain fixed on the shaft. At the neutral axis they are at 45 degrees to the shaft axis, as given by torsion and shear. At the points of highest tensile or compressive bending stress they are rotated to align more closely with the bending stresses as shown in Fig. 9.2. Sines' theory works because the principal axes of the alternating stresses remain fixed and the sum of mean normal stresses is invariant, regardless of the choice of orthogonal axes.

Self-stresses must, of course, be considered with the mean stresses. Our examples assumed the absence of self-stresses.

For in-phase stresses the use of equivalent stress equations is straight-forward. For out-of-phase biaxial stresses graphical methods are preferable. The ellipses of constant octahedral stress can conveniently be drawn because their proportions are those of isometric ellipses, for which templates are available.

According to Sines' method mean torsion stress has no direct influence on fatigue. It may produce yielding and thus change self-stresses and indirectly affect fatigue resistance. Conventional data sheets on helical springs seem to contradict this. They are drawn like Goodman diagrams with straight lines starting from the range for zero minimum stress to a point of equal minimum and maximum stress.

A leading springmaker confirmed that constant stress ranges up to near yielding would represent test data more closely. The conventional ''Goodman'' shape is intended as a simplified safe first approximation.

Sines' method can be extended to use strain instead of stress as the independent variable. The equivalent alternating strain ϵ_q (corresponding to $\Delta\epsilon/2$ is

$$\epsilon_q = \frac{[(\epsilon_{a1} - \epsilon_{a2})^2 + (\epsilon_{a2} - \epsilon_{a3})^2 + (\epsilon_{a3} - \epsilon_{a1})^2]^{1/2}}{(1 - \nu)\sqrt{2}} \qquad (9.9)$$

where ϵ_{a1} is an alternating principal strain and ν is Poisson's ratio, which changes from about 0.3 for elastic behavior to 0.5 for purely plastic behavior.

The method can then be used for low cycle fatigue by entering an $\epsilon - N$

curve with ϵ_q. However, there is no experimental confirmation for this extension of Sines' method.

9.2.2 Langer's Method

Langer's method [8] uses the maximum alternating shear stress as failure criterion. If the alternating stresses are produced by a single alternating load the maximum alternating shear stress can be determined from the maximum and minimum load. The expected fatigue life is then obtained from a uniaxial $S-N$ curve, for the applicable mean normal stress, by using an equivalent stress

$$S_a = 2\tau_a \qquad (9.10)$$

Langer's method can also be extended to use strain instead of stress as input.

Kelly [9] describes the application of Langer's method to nonproportional loading in some detail. It is handled in the same manner which is now explained for the computation of an equivalent shear strain.

The equivalent alternating strain ϵ_q is computed from the maximum alternating shear strain γ_a

$$\epsilon_q = \frac{2\,\gamma_a}{(1 + \nu)} \qquad (9.11)$$

where ν is Poisson's ratio.

The value of γ_a is computed from the principal shear strains:

$$\gamma_{12} = \frac{(\epsilon_1 - \epsilon_2)}{2}$$

$$\gamma_{23} = \frac{(\epsilon_1 - \epsilon_3)}{2} \qquad (9.12)$$

$$\gamma_{31} = \frac{(\epsilon_3 - \epsilon_1)}{2}$$

They are computed at two instants during the cycle chosen in such a way that the maximum γ_a is obtained from the relation

$$\gamma_a = (\gamma_{12} \text{ at instant } \#1) - (\gamma_{12} \text{ at instant } \#2) \qquad (9.13a)$$

or

$$\gamma_a = (\gamma_{23} \text{ at instant } \#1) - (\gamma_{23} \text{ at instant } \#2) \qquad (9.13b)$$

or

$$\gamma_a = (\gamma_{31} \text{ at instant } \#1) - (\gamma_{31} \text{ at instant } \#2) \qquad (9.13c)$$

whichever is the greatest in absolute magnitude.

9.3 COMPLEX MULTIAXIAL STRESSES

In many cases the principal directions of the alternating stresses are not fixed in the stressed part but change orientation. Crankshafts are a typical example. Pipes subjected to out-of-phase torsion and bending are another. Methods for handling such situations have been developed by the pressure vessel and piping industry. We now discuss Garud's method [10] and an extension of Langer's method [8, 9].

9.3.1 Garud's Method

Garud uses the plastic work per cycle as parameter for life to crack initiation. Plastic work is calculated by integrating the product of stress times plastic strain increment (the area of the hysteresis loop) for each of the six components of the stress tensor. The sum of the six integrals is the plastic work per cycle. The determination of the hysteresis loops $\sigma_x - \epsilon_x$, $\tau_{xy} - \gamma_{xy}$ and so on is the difficult part of this method. It requires careful consideration of hardening rules and flow rules, as described in Garud's paper [10]. As an example of the hysteresis loops in complex loading we show loops for axial stress–strain and for torsional stress–strain in Fig. 9.3 for a load cycle consisting of torsion and bending, with $\gamma = 1.5\epsilon$ and bending leading torsion by 45 degrees. Figure 9.4 compares lives from fatigue tests with out-of-phase bending and torison by Kanazawa et al. [11] with plastic work per cycle per unit volume.

Although the correlation is rather good, plastic work in torsion produces only about half the fatigue damage of plastic work in bending. Secondary

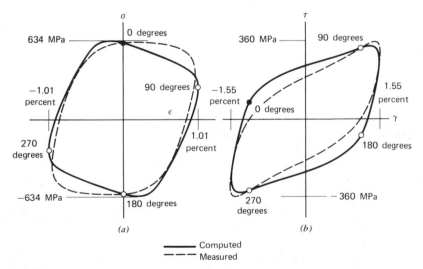

FIGURE 9.3 Stress and strain response for 45 degree out-of-phase bending and torsion [10]. (*a*) Bending. (*b*) Torsion.

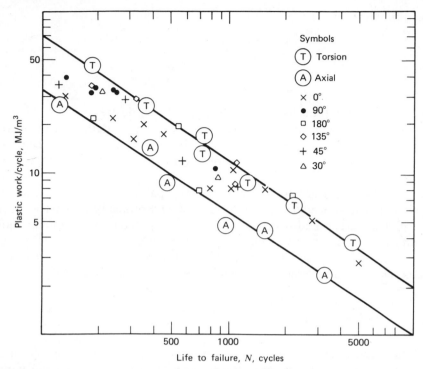

FIGURE 9.4 Plastic work per cycle (calculated) versus life to failure (observed) for combinations of torsion and bending at various phase angles [11] (reprinted with permission of the American Society of Mechanical Engineers).

effects caused by normal stresses may be introduced to account for the difference. This has not yet been done.

Garud's method accounts for an observation [11] that is contrary to the results of the other methods: At low stresses in-phase loading is more dangerous than out-of-phase loading. At stresses well above the yield strength out-of-phase loading can be more dangerous.

9.3.2 Langer's Method for Moving Principal Axes

The extension of Langer's method to general multiaxial loading, without restrictions as to fixed principal axes, uses the six components of the stress tensor or of the strain tensor. They are σ_x, σ_y, σ_z, τ_{xy}, τ_{yz}, τ_{zx} and the corresponding strains ϵ and γ.

Two instants during the cycle are chosen. We call them b for beginning and f for final. The stress values at those instants are

$$\sigma_{xb} \quad \text{and} \quad \sigma_{xf}$$

$$\sigma_{yb} \quad \text{and} \quad \sigma_{yf}$$

$$\text{etc.}$$

The differences

$$\delta_x = \sigma_{xb} - \sigma_{xf}$$

$$\delta_y = \sigma_{yb} - \sigma_{yf}$$

etc.

are formed. Kelly [9] calls these differences "pseudostresses."

The principal values of the pseudostresses are computed. Half the difference between principal pseudostresses is the pseudo-shear stress. That pair of instants b and f is sought that produces the maximum pseudo-shear stress.

It may be necessary to investigate many pairs of instants to find the maximum. The procedure is slightly easier, and probably at least as accurate, if octahedral pseudo-shear stresses are computed rather than maximum pseudo-shear stresses, as in one case of the ASME boiler code [12]. Test data [10, 11] show that this method can underestimate the danger of 90 degree out-of-phase torsion and bending at short fatigue lives.

9.4 CRACK PROPAGATION

In a first approximation, the propagation of cracks depends on the alternating stresses that pull the crack faces open; the other principal alternating stresses have a modifying influence. In this section we do not deal with the rate of crack propagation in multiaxial stress fields, but only with the condition for zero crack growth. A small alternating tensile stress is required for crack growth.

In the stress plane for biaxial states of stress, the compression–compression quadrant is entirely safe from crack growth. Any state of nominal mean and alternating stresses represented by a line that remains entirely in this quadrant will not propagate small cracks.

A small band of tensile stress, corresponding to the critical alternating tensile stresses, S_{cat}, can be added to this quadrant as shown in Fig. 9.5.

If the mean stress is far in the tension–tension quadrant, the zone of stress states that leads to nonpropagating cracks is a small square.

Nominal net stresses are the criterion for crack propagation. They must, of course, include self-stresses. Local stresses are the criterion for crack initiation. Therefore, the initiation ellipse can become smaller than the propagation rectangle if stress concentrations are present.

All these considerations for long life apply only if gross yielding is avoided. The zone of nonyield is an ellipse according to the von Mises–Huber–Henky or octahedral shear stress theory, or a hexagon according to the Tresca theory.

In Fig. 9.6 this approach is applied to the determination of the optimum amount of surface compression (shot-peening followed by partial stress

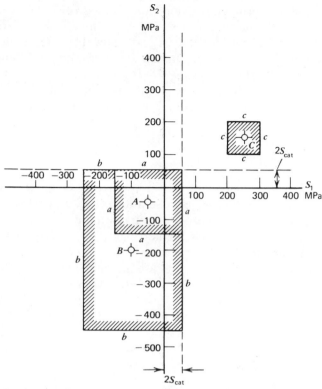

FIGURE 9.5 Areas of "safe" alternating tensile stress smaller than S_{cat}. Mean stresses A, B, C. Boundaries of corresponding areas of safe alternating stresses aa, bb, cc. Boundaries are symmetrical about the mean stress points.

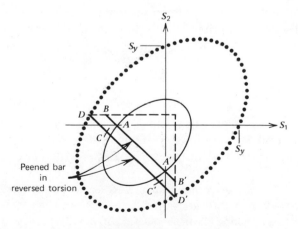

FIGURE 9.6 Three criteria used for optimizing a biaxial stress state. (——) Crack initiation ellipse, (- - -) crack propagation lines, (····) yield ellipse.

relief baking) for coil springs or torsion bars. For a given compressive mean, line $A-A'$ represents the torsional stress range for crack initiation, while $B-B'$ is that for propagation. The largest allowable alternating stress is achieved for a mean stress where failure due to propagation and yielding are equally likely, as shown by line $D-D'$. In this, $A-A'$ shifts to $C-C'$ (larger ellipse). Further increases in compressive mean stress would reduce allowable alternating stress because the latter is limited by yielding.

9.5 CUMULATIVE DAMAGE

Very little is known about methods for predicting cumulative damage in multiaxial fatigue. Nevertheless, excellent axles and crankshafts subject to complex multiaxial fatigue are produced, based on experience.

We have seen that in Langer's method one must search for the instants that produce maximum differences. These instances correspond to the peaks and valleys, or reversal points, which are an elementary input for uniaxial cumulative damage calculations. Jones et al. [13] propose a cycle counting procedure for multiaxial stress histories. Garud's method [10] can be adapted by calculating damage from plastic work per cycle.

9.6 DOS AND DON'TS IN DESIGN

1. Do check whether the alternating stresses or strains have fixed principal directions. If so, fairly simple methods can be used.
2. Don't assume that the stresses at notch roots are of the same type (uniaxial or multiaxial) as the nominal stresses. The stresses at a notch in a tensile bar are multiaxial.
3. Do remember that yielding, crack initiation, and crack propagation are governed by different criteria, with different equivalent stresses.
4. Do try to learn from experience with similar parts similarly loaded.

REFERENCES FOR CHAPTER 9

1. A. Nadai, *Theory of Flow and Fracture of Solids,* McGraw-Hill Book Co., New York, 1950.
2. G. E. Dieter, Jr., *Mechanical Metallurgy,* McGraw-Hill Book Co., New York, 1961.
3. R. vonMises, "Mechanik der festen Koerper in plastischdeformablen Zustand," *Goettinger Nachrichten,* 1913.
4. H. Henky, "Zur Theorie Plastischer Deformationen," *Z. Angew. Math. Mech.,* 1924.

5. M. I. Huber, in *Theory of Flow and Fracture of Solids,* McGraw-Hill Book Co., New York, 1950, p. 230.

6. V. V. Novozhilov, *Appl. Math. Mech. (USSR),* Vol. 16, No. 5, 1952, p. 617.

7. G. Sines, "Failure of Metals Under Combined Repeated Stresses With Superimposed Static Stresses," NACA Tech. Note 3495, 1955; also *Elasticity and Strength*, Allyn and Bacon, Boston, MA 1960.

8. B. F. Langer, "Design of Pressure Vessels Involving Fatigue," *Pressure Vessel Engineering,* R. W. Nichols Ed., Elsevier Publishing Co., Amsterdam 1971.

9. F. S. Kelly, "A General Fatigue Evaluation Method," ASME preprint 79-PVP-77, 1979.

10. Y. S. Garud, "A New Approach to the Evaluation of Fatigue under Multi-axial Loadings," *Methods for Predicting Material Life in Fatigue,* W. J. Ostergren, Ed., ASME, 1980, p. 247.

11. K. Kanazawa, K. J. Miller, and M. W. Brown, "Low Cycle Fatigue Under Out-of-Phase Loading Conditions," *Trans. ASME, J. Mater. Eng. Technol.,* Vol. 99, 1977, p. 22.

12. *Cases of ASME Boiler and Pressure Vessel Code,* 1974, Case 1592.

13. D. P. Jones, C. M. Friedrich, and R. G. Hoppe, "A Cycle Counting Procedure for Fatigue Failure Predictions for Complicated Multi-Axial Stress Histories, ASME preprint 77-PVP-29, 1977.

PROBLEMS FOR CHAPTER 9

1. An AISI 8640 steel shaft is Q & T to S_u = 1400 MPa (200 ksi) and S_y = 1250 MPa (180 ksi) with RA = 10 percent. A step exists in the shaft that causes K_t = 2.1 in torsion. If the smaller shaft diameter is 50 mm, what maximum pulsating torque can be applied over a period of several years and last 10 million cycles?

2. Repeat problem 1 if the shaft is shot-peened, causing compressive self-stresses at the step equal to 700 MPa in both the longitudinal and transverse direction.

3. In problems 1 and 2, what maximum torque can be applied for 50,000 cycles?

4. A 25 mm diameter 2024-T3 aluminum shaft is subject to in-phase bending and torsion. The bending moment varies from 0 to 144 N·m, while the torque varies from −100 to +200 N·m. Will the shaft withstand 10 million cycles of this combined loading?

5. For problem 4, how much could the bending moment be increased and still have a life of 50,000 cycles of combined loading?

CHAPTER **10**

FATIGUE FROM REAL LOAD HISTORIES

Real load histories are usually irregular as shown in Figs. 5.2a to 5.2c. The loads on an aircraft wing or on a tractor shovel are far from uniform. Techniques of analysis and of testing have been developed to predict whether such loads will produce acceptable or unacceptable fatigue lives. Data from constant amplitude tests are the basis for the analysis. The analysis may be simple, based on nominal stresses and the assumption that damage is linear with the number of cycles. Or it may be more complex, for instance, in considering the early stages of fatigue by notch strain analysis [1] and the later stages by crack growth analysis. Which approach is most appropriate in a particular instance depends on several criteria, which are discussed in Section 10.5.

10.1 CUMULATIVE DAMAGE

In tests with uniform load cycles the cumulative effect of all the cycles eventually produces fatigue failure (unless the load is below the prevailing fatigue limit). When the loading and unloading does not occur in uniform cycles but in an irregular manner, the cumulative effect of these events may also produce fatigue failure. The term "cumulative damage" refers to the fatigue effects of loading events other than uniform cycles. The term "spectrum" as used in fatigue literature often means a series of fatigue loading events other than uniformly repeated cycles. Sometimes spectrum means a listing, ordered by size, of components of irregular sequences, as, for instance, in Table 10.1. Other parameters, such as maximum load, or pairs of parameters, such as range and mean, are also used to define the classifications or "boxes" in which the counts of cycles are listed.

Figure 10.1 shows three service load records [2] along with details to a larger time scale of spectrum b as examples of irregular real load histories. One approach to variable load histories uses the concept of damage, defined as the fraction of life used up by an event or series of events. These fractions

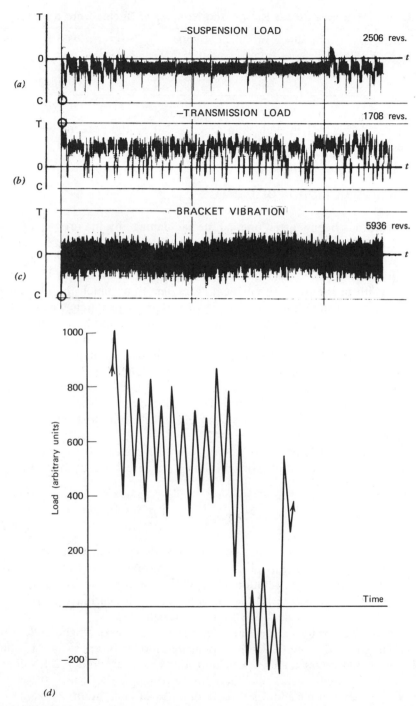

FIGURE 10.1 Amplitude–time display of service histories [1] (reprinted with permission of the Society of Automotive Engineers).

TABLE 10.1 Example Stress Range and Number of Cycles from a History

Stress range, ΔS, (MPa)	<300	300	500	700	900	1100
Number of cycles		723	312	0	19	3

are added together; when their sum reaches 1.0 or 100 percent we expect and predict failure.

10.2 THE USEFUL FICTION OF DAMAGE

After a crack has been started one can define damage by the growth of that crack. According to this definition loading events that produce zero crack growth would do zero damage. A loading event that extends the crack by $\Delta a = 0.01$ mm would do damage, which might be defined as $\Delta a/W$, where W is the width of the part. The critical crack length at which the next high load will produce fracture might be as small as $0.5W$. The damage calculated to that point would be only 50 percent, yet fracture would be imminent. A more satisfactory definition of damage might be $D = \Delta a/a_c$, where a_c is the critical crack length for the highest expected load. Actually one usually avoids the concept of damage in crack growth calculations. One generally works with da/dN, sums the crack increments, and arrives at a critical crack length and prediction of failure. The same event that produced 0.01 mm crack growth at an earlier stage may produce 1 mm crack growth at a later stage.

In earlier stages of fatigue one cannot refer to a measurable quantity like crack length as a measure of damage. The damage caused by one cycle is defined as

$$D = \frac{1}{N}$$

where N is the number of repetitions of this same cycle that equals the median life to failure. Figure 10.2 shows an $\epsilon - N$ curve and the corresponding $D - \epsilon$ curve. The relation $D = 1/N$ expresses the linear damage rule, proposed by Palmgren [3] for prediction of ball bearing life and later again by Miner [4] for prediction of aircraft fatigue life.

Applied to ball bearings the linear damage rule works in the following way: Tests show the median life of bearings model X operating at 3000 RPM to be 1100 hours under 1 kN load and 150 hours under 2 kN load. How many hours can we expect the bearing to last if the load is 1 kN for 55 minutes of each hour and 2 kN for 5 minutes of each hour?

If the damage done in 1 hour or 60 minutes is D_{60}, then the life in hours will be $1/D_{60}$. The damage done in each hour or 60 minutes will be the sum of the damages done by the 2 kN load for 5 minutes plus the damage done by the 1 kN load for 55 minutes. Each of these is equal to the fraction of (time at load/time to failure)

$$D_{60} = \frac{5 \text{ minutes}}{150 \times 60 \text{ minutes}} + \frac{55 \text{ minutes}}{1100 \times 60 \text{ minutes}}$$

$$= 5.55 \times 10^{-4} + 8.33 \times 10^{-4} = 13.88 \times 10^{-4}$$

and the expected life is $1/D_{60} = 1/(13.88 \times 10^{-4}) = 720$ hours.

Although the assumption of linear damage is open to many objections it

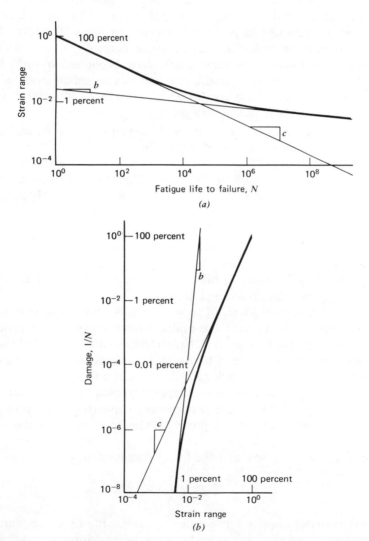

FIGURE 10.2 Fatigue test data in terms of (a) life and (b) linearlly calculated damage.

is used because none of the other proposed methods achieves better agreement with data from many different tests. Below are some of the objections one can raise against the linear damage rule.

1. Load histories may be irregular. How to convert irregular histories to an equivalent sum of cycles of various magnitudes is discussed in Section 10.4.
2. The sequence of events may have a major influence on the fatigue life. This is discussed in Section 10.3.
3. The life N is not clearly defined. Does it mean fracture or loss of stiffness? Is it obtained by repeated applications of the same loads or of the same deflections? In general N means the life to fracture of small smooth specimens tested in load or strain control. In such specimens fracture follows soon after small cracks have produced loss of stiffness. In some cases N may refer to life of a part to a specified mode of failure under some particular test conditions, for instance, 20 percent load decrease under uniform deflection cycles.
4. There is no reason to believe that the damaging effect of the eighth cycle is the same as that of the $(N - 8)$th cycle.

Palmgren and Miner were well aware of this. They decided to use an average damage and their method is still the best we have. In the literature the damage produced by n cycles, which is

$$nD = \frac{n}{N} \tag{10.1}$$

is often called "cycle ratio," which appears to avoid the implication that we can actually calculate fractional damage.

There have been attempts [5, 6] to work without the linear damage rule. These proposed methods treat the entire fatigue process as a process of crack propagation. They define a "characteristic material defect length" as a materials property. The size of the small initial crack is chosen so that by applying the recommended crack growth law to it one arrives at the known fatigue life. Another, more successful, approach uses the linear damage law for the early stage of the fatigue process and computes crack propagation for the later stage when cracks grow by tensile stresses in the opening mode.

In summary, we may say that the linear damage rule

$$D = \frac{1}{N}$$

is the best available approach to predicting fatigue life to the appearance of cracks that open and close and in many cases the best approach to predicting the total fatigue life to fracture.

10.3 HISTORIES WITH AND WITHOUT SEQUENCE EFFECTS

Sequence effects exist both in the early stages (crack initiation) and in the later stages (crack propagation) of fatigue. The same principles govern both stages. We begin by looking at the early stages.

It has been shown [7] that fatigue strength of smooth specimens is reduced more than indicated by the linear damage rule if a few cycles of high stress are applied before testing with lower stresses. This effect, however, is very small compared to sequence effects on notched parts. On notched parts the sequence effect can be very strong. We recall Table 7.1, which shows that the fatigue strength of notched specimens could be more than doubled by a single high overload. The fatigue life might be 10 times or even 100 times as great if that overload is applied at the beginning of the sequence rather than at the end of the sequence. We also recall from Chapter 7 that this effect could be explained by self-stress effects.

Figure 10.3, taken from Crews [8], shows another sequence effect. Here the life of a specimen with a hole was 460,000 cycles of low load after 9.5 cycles of high load but only 63,000 cycles after 10 cycles of high load. This difference is also explained by the self-stresses remaining from the high loads. When the high load cycles ended on a tension peak the effect was beneficial to fatigue life; when they ended on a compression reversal the effect was harmful.

In Figs. 10.4 and 10.5, reprinted from Stephens et al. [9], sequence effects are shown in terms of crack propagation rates da/dN. They confirm the results that were given above in terms of life to failure N. Yielding and the resulting self-stresses and crack closure near the crack tip are the main causes for these effects.

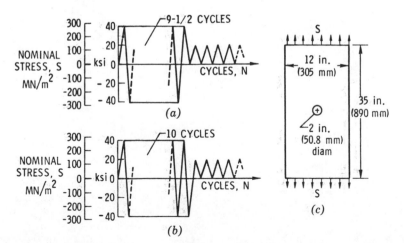

FIGURE 10.3 Effect of prior loading on fatigue life [8]. (*a*) Beneficial prior loading. (*b*) Detrimental prior loading. (*c*) Notched specimen.

Here, even more than in the early stages of fatigue, the details of the sequence, the directions of the last overloads, are of prime importance. Even a small tensile load can produce a plastic zone at the tip of the crack, forming compressive self-stresses that retard the growth of cracks. Compressive loads act differently. They do not open the cracks and they must be of greater magnitude before they produce yielding. Then they form tensile self-stresses at the crack tip and thus accelerate the growth of cracks.

We see that sequence effects can be very important and that they depend not on number of cycles but on the exact details of the load history. This requires, of course, detailed step-by-step analysis. Many service histories, however, are such that sequence effects either cancel each other or are entirely unpredictable. In those cases nothing is gained by the detailed analysis, and the shorter prediction methods should be used because they provide equally close predictions and do not create illusions of higher accuracy. We do not yet have quantitative rules that tell when sequence effects must be considered in predicting fatigue life. A few qualitative rules based on limited experience can be stated as follows:

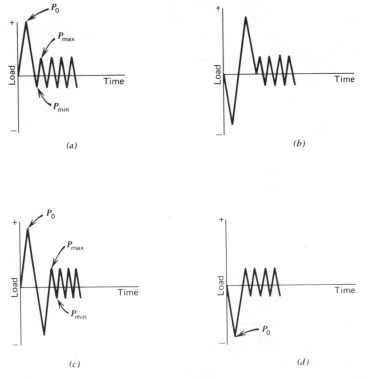

FIGURE 10.4 Four different overload patterns: (a) tension, (b) compression–tension, (c) tension–compression, (d) compression.

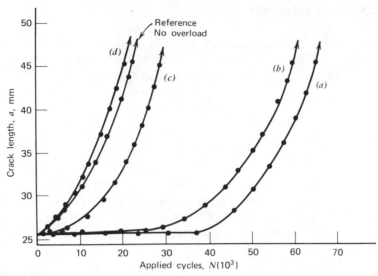

FIGURE 10.5   Crack growth following different overload patterns in 7075-T6 [9] (reprinted by permission of the American Society for Testing and Materials). (*a*) Tension overload, (*b*) compression–tension overload, (*c*) tension–compression overload, (*d*) compression overload. $K_{ol} = 45.3$ MPa$\sqrt{m}$. $K_{max} = 19.7$ MPa$\sqrt{m}$.

1. If the sequence of service loads is completely unknown one must decide whether to assume significant sequences or not.
2. If the loading is Gaussian random with a narrow frequency band there will be no definable sequence. Figure 10.1c is an example.
3. If the loading history shows infrequent one-sided spikes, as for instance in the ground–air–ground cycle of aircraft, one should expect sequence effects.
4. Infrequent tensile overloads produce retardation of crack growth or crack arrest. Compressive overloads large enough to produce yielding can produce the opposite effect.

For the histories shown in Fig. 10.1 predictions that included sequence effects were not significantly better than predictions that neglected sequence effects [10, 11]. The negligible influence of sequence is explained by the short intervals between the large amplitudes, which produce by far the greatest damage. If we had fewer large amplitudes and many more small amplitudes (as in aircraft wings), the damage done by the small amplitudes, and the sequence effects, would be significant.

In the preceding discussion "significant" means larger than the uncertainty inherent in our calculations.

10.4   COUNTING METHODS

The object of all cycle counting methods is to compare the effect of irregular load histories to $S-N$ or $\epsilon-N$ curves obtained with uniformly repeated simple load cycles. The application of the linear damage rule

$$D = \frac{1}{N}$$

requires that we know the condition (mean and amplitude of stress or strain) to which the damaging event should be compared. Different counting methods can change the resulting predictions by an order of magnitude [12]. A very simple example is shown in Fig. 10.6. The history in Fig. 10.6a could be analyzed as cycles as shown in 10.6b, which uses only the segment from point 1 to point 11, which is the beginning of a repetition of the same events. In this method of counting we find one each of cycles with the following five pairs of extreme values:

$$100/200 \qquad 100/300 \qquad -200/+200 \qquad -200/-100 \qquad -300/-100$$

Another counting method produces the five pairs shown in 10.6c:

$$-300/300 \qquad 100/200 \text{ twice} \qquad -200/-100 \text{ twice}$$

If damage is proportional to the sixth power of the ranges we calculate for Fig. 10.6b a damage proportional to

$$2 \times 1^6 + 2 \times 2^6 + 4^6 = 2 + 132 + 4096 = 4230$$

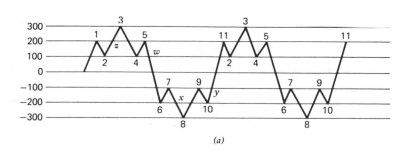

(a)

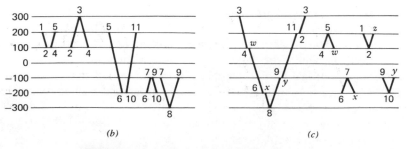

(b)   (c)

FIGURE 10.6   Two ways of counting cycles.

and for Fig. 10.6c a damage proportional to

$$6^6 + 2 \times 1^6 + 2 \times 1^6 = 46{,}656 + 2 + 2 = 46{,}660$$

The latter method calculates 11 times as much damage as the former, and the results of the latter method correlate much better with test results.

All good counting methods must count a cycle with the range from the highest peak to the lowest valley and seek to count other cycles in a manner than maximizes the ranges that are counted. This rule can be justified either by assuming that damage is a function of the magnitude of the hysteresis loop or by considering that in fatigue (as in many other fields) intermediate fluctuations are less important than the overall differences between high points and low points.

All good counting methods count every part of every overall range once and only once. They also count smaller ranges down to some predetermined threshold once and only once. Three counting methods that achieve this objective are well documented in the literature: range-pair, rainflow, and racetrack.

10.4.1 Range-Pair Method

Figure 10.7 shows the operation of this method [13, 14]. Small cycles are counted first and their reversal points (peaks and valleys) are eliminated from further consideration. In Fig. 10.7a there are 20 reversals. Fourteen of them are counted and eliminated by counting the seven pairs indicated by cross hatching. This leaves the 6 reversals of Fig. 10.7b. Looking only at Fig. 10.7b we see three peaks and three valleys as follows:

```
      250        140         160          (0)
  (0)        50        -140       -120
```

They would be counted as cycles (shown by maximum and minimum loads separated by a slant as follows:

```
   50/140 (cross hatched)     160/-120 (cross hatched)
```

leaving 250/-140, which is shown in Fig. 10.7c. The result of a range-pair count is a table of the occurrence of ranges and, if desired, of their mean values.

10.4.2 Rainflow Method

The operation of the rainflow method is shown in Fig. 10.8 for a history consisting of four peaks and four valleys. The rules are:

1. Rearrange the history to start with the highest peak.
2. Starting from the highest peak go down to the next reversal. Proceed horizontally to the next downward range; if there is no range going down

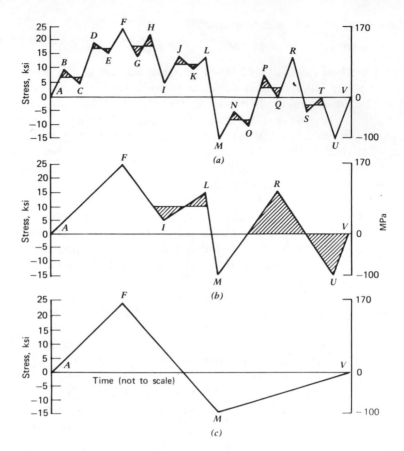

FIGURE 10.7 Hayes' method of counting from E. F. Bruhn, *Analysis and Design of Flight Vehicle Structures* [13, 14].

from the level of the valley at which you have stopped, go upward to the next reversal.

3. Repeat the same procedure upward instead of downward and continue these steps to the end.

4. Repeat the procedure for all the ranges and parts of a range that were not used in previous procedures.

   This assumes that the highest peak is more extreme than the lowest valley. If not, start with the lowest value and go up instead of down.

In Fig. 10.8a the first traverse is shown. Remaining ranges are in Fig. 10.8b. The procedure applied to 10.8b again leaves a pair of ranges unused. They are shown in 10.8c. A result of this count might be a list of cycles:

$$25/-14 \qquad 14/5 \qquad 16/-12 \qquad 7/2$$

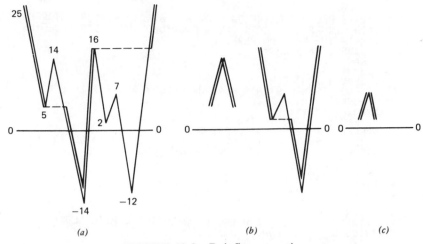

(a)   (b)   (c)

FIGURE 10.8 Rainflow counting.

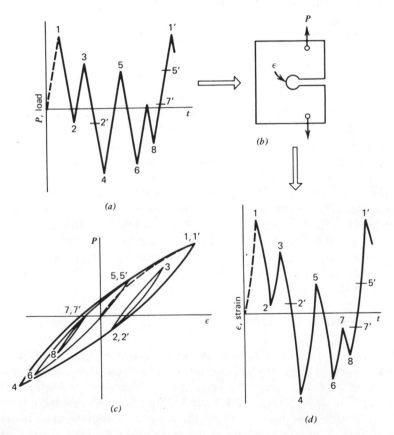

(a)

(b)

(c)

(d)

FIGURE 10.9 A load versus time history repeatedly applied to a notched component and the resulting notch strain response [15] (reprinted with permission of the Society of Automotive Engineers).

```
C******THIS PROGRAM RAINFLOW COUNTS A STRAIN HISTORY STORED
C******IN SEQUENCE ON DATA CARDS PROVIDED THE HISTORY BEGINS
C******WITH THE MAXIMUM PEAK.
        DIMENSION E(25)
        N=0
100     N=N+1
        READ(1,200) E(N)
C******READ NOMINAL STRAIN PEAK OR VALLEY AND PUT IN E(N)
200     FORMAT(F10.0)
C******POINT A*****
300     IF(N.LT.3) GO TO 100
C******MAKE SURE E HAS AT LEAST 3 POINTS
        IF(ABS(E(N)-E(N-1)).LT.ABS(E(N-1)-E(N-2))) GO TO 100
C******CHECK TO SEE IF ANY LOOPS HAVE CLOSED
C******IF NOT, GO BACK TO 100
        RANGE=ABS(E(N-1)-E(N-2))
C******CALCULATE STRAIN RANGE
C******POINT B*****
        N=N-2
        E(N)=E(N+2)
C******THROW AWAY PEAK AND VALLEY ASSOCIATED WITH CLOSED
C******LOOPS, AS NO LONGER NEEDED.
        IF(N.NE.1) GO TO 300
C******CHECK TO SEE IF ALL LOOPS ARE CLOSED
C******IF NOT, GO BACK TO 300
        STOP
        END
```

FIGURE 10.10   Cycle counting algorithm [16].

which is the same result as that found by range-pair count except for the additional small cycle 7/2, which was added to the history.

The advantage of rainflow counting comes when it is combined with a strain analysis as shown in Fig. 10.9 [15]. The damage can then be computed for each cycle as soon as it has been identified in the counting procedure and the corresponding reversal points can be discarded. A computer program that accomplishes this counting is shown in Fig. 10.10 [16]. Other methods that accomplish the same purpose have been shown by a number of authors [1, 17]. The procedure is called "rainflow counting" because it was first worked out by M. Matsuishi and T. Endo [18], who thought of the lines going horizontally from a reversal to a succeeding range as rain flowing down a pagoda roof represented by the history of peaks and valleys after it is rotated 90 degrees.

10.4.3   Racetrack Counting

The racetrack method of counting cycles is shown in Fig. 10.11 [13]. The original history in Fig. 10.11a is condensed to the history in Fig. 10.11c. The method of eliminating smaller ranges is indicated in Fig. 10.11b. A "racetrack" or width S is defined, bounded by "fences" that have the same profile as the original history. Only those reversal points are counted at which a "racer" would have to change from upward to downward, as at f and n, or vice versa as at m and o. The width S of the track determines the number of reversals that will be counted.

This method was originally called the "ordered overall range method." Its object is to condense a long complex history of reversals, or a long complex chart of peaks and valleys, to make it more useful. The condensed listing can be used as a record of the most essential features of the chart or of the history, and as a basis for calculations, forecasts, or characterizations.

The rationale behind the method is as follows: The distance from the highest peak to the lowest valley is the most important feature of the history.

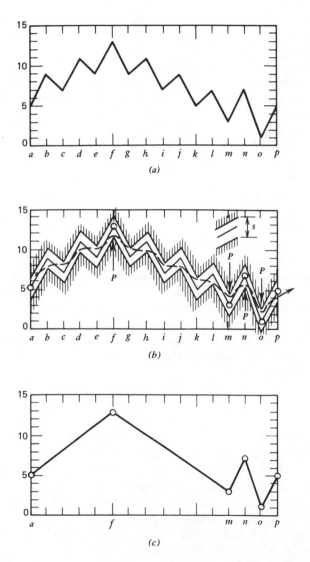

FIGURE 10.11 The racetrack method of listing reversals [13]. (a) An irregular history. (b) Screening through width S. (c) The resulting condensed history.

The distance from the second lowest valley to the second highest peak is the next most important feature, provided that the second range (second valley to second peak) crosses the first range (maximum peak to maximum valley), or is outside of the time interval defined by the first two. Continuing in this manner one can either exhaust the count of all the reversals or stop counting at a selected point and consider all the lesser peaks and valleys as nonessential and negligible.

The racetrack method can be used in two modes. The first produces a histogram, spectrum, or listing in which magnitudes of overall ranges and frequencies of their occurrences are given. (The distance from a high peak to a low valley is called the overall range between the two.) If one continues listing until all peaks and valleys have been counted, this mode produces the same results as the rainflow method and almost exactly the same results as the range-pair method. With the racetrack method one can stop counting after only a fraction of all peaks and valleys have been listed. In many cases this leads to the same result as if all reversals had been listed.

The second mode produces a condensed history in which essential peaks and valleys are listed in their original sequence. If one continues this mode of listing until all reversals have been listed, then the entire history will be reproduced in all its detail. If one stops before all reversals are listed, one omits detailed features of the history but retains the reversals that produce the largest overall ranges.

The condensed history contains more information than the spectrum. It includes the sequence of events, which may be important if yielding produces residual stresses that remain active for many succeeding reversals.

This method picks out major overall ranges first, and keeps them in the correct time sequence. It is useful for condensing histories to those few events–perhaps 10 percent of all—that do most of the damage—usually more than 90 percent of all calculated damage [12]. The condensed histories accelerate testing and computation and permit focusing of attention on a few significant events. A computer program for the racetrack method was published by Fuchs et al. [12].

All three methods—range pair, rainflow, and racetrack—can achieve the same objective of listing overall ranges and interruptions of overall ranges in order of magnitude with or without mean stress. The rainflow method is currently most popular. Devices for real time rainflow counting of strain gage signals are commercially available. Rainflow counting is most advantageous when used for analysis of a notch–strain history. Note that the rainflow count of load reversals may differ from the rainflow count of strain reversals. For notch strain analysis the load history is first converted to a notch–strain history and then rainflow counted. Racetrack counting is most suitable for condensing histories. It is usually applied to the load history, which is the same as a history of elastic strains.

10.4.4   Other Methods

Cycle counting is not the only method that can be used for assessing cummulative damage. Statistical measures of the events in a measured or expected load history may be useful. It has been shown [19] that the damage produced by random loadings is proportional to the "root-mean sixth power" of the load ranges. In a statistical approach the details, which may be all-important in fatigue analysis, are submerged. The root-mean sixth power is mentioned less to recommend it than to discourage the use of root-mean-square in general fatigue analysis. Root-mean-square measures have been used successfully for predicting crack growth with the spectrum loading in Fig. 5.2c [20].

10.5   PREDICTING THE EFFECTS OF IRREGULAR LOAD HISTORIES

When the future load history is known one can follow it reversal by reversal, either by test programs applied to the part or analytically [10, 17, 21]. Less sophisticated methods may achieve equally accurate forecasts, depending on the certainty with which the future load history and the fatigue properties of the parts are known. In general the future load history is not known precisely. Therefore it would make little sense to try to allow for sequence effects and other interactions of the loading events. In some cases either the future load history can be predicted fairly well or it is prescribed by the customer as the basis for the analysis and for acceptance tests.

   If sequence effects can be neglected the predictive calculations can be simpler. Sequence effects must be expected when between many minor ranges there are a few major deviations that always end in the same way, coming back to the minor ranges either always from high compression or always from high tension.

   Figure 10.1b shows a history that could be expected to show sequence effects because between the 1600 minor ranges there are about 100 major deviations that always come back from high compression. Calculations with and without considering sequence effects were made for two materials and three load levels and these were compared with fatigue test results for a notched specimen [17]. There were no significant differences for the softer material, Man-Ten, nor for the lowest load level, which produced fatigue lives of about 40,000 blocks of 1708 reversals. For the harder material, RQC-100, at higher load levels the results were as listed below.

Test lives	22–30	269–460	Blocks
Calculated without sequence	69	1300	Blocks
Calculated with sequence	10	170	Blocks

The numbers quoted above are for lives to the appearance of obvious cracks (2.5 mm long).

The more sophisticated methods of calculation require knowledge of the monotonic and cyclic stress–strain curves in addition to the fatigue properties. Values of S_y', K', and n' are given in Table A.2 for selected engineering alloys. Socie [17], however, has shown widely divergent values quoted for samples of one lot of steel.

If a proven computer program is available the more sophisticated methods are not much more difficult to execute than the simpler methods. If no proven program is available the simpler methods have great advantages in ease of verification and possibly in economy of costs.

When the questions implied above have been answered one proceeds to the chosen method of calculation. None is perfect. Of the many that have been proposed we discuss one simpler method and a few more sophisticated approaches.

10.5.1 A Simple Method of Life Prediction

This method neglects sequence effects. Often (but not always) it produces predictions that are just as accurate as those of more complex methods when compared to test results.

An S–N curve is used as primary input. If an experimental load versus life curve of the part is available, it is the best input. If an experimental S–N curve of the material is available, it must be modified to account for self-stresses and stress concentrations at the critical location, as is explained in Chapter 8. The full basic notch factor K_b is applied at 10^7 cycles. No notch factor is applied at 1 cycle. These two points are joined by a straight line on log–log coordinates. An estimated notched S–N curve in terms of alternating nominal stress versus cycles to failure is thus obtained. For the most preliminary of estimates it may even be necessary to estimate an S–N curve for the material, as is discussed in Chapter 8.

The next step is to count load ranges of the given load history by rainflow or racetrack counting as is discussed in Section 10.4. In most cases the 10 percent of all overall ranges that are greatest will do more than 90 percent of the damage [10, 12]. These greatest overall ranges can easily be picked out either from a computer printout or by eye from a plot of peaks and valleys. Figure 10.12 shows how to do this after making two masks of the load history, neglecting obviously minor wiggles.

Next the load ranges must be converted to nominal stress ranges. Finally the damage expected from each of the stress ranges is calculated from the S–N curve as $1/N$, and the damages are added. The ratio of 1/(sum of damages) is the number of times we expect the given history to be endured until failure occurs.

10.5.2 Sophisticated Methods of Life Prediction

All the more sophisticated methods treat the early stages (crack initiation) separately from the later stages (crack propagation). The transition from

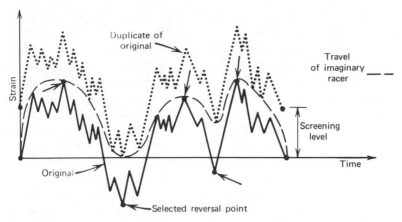

FIGURE 10.12   How the ordered overall range method, using the "racetrack" analogy, can be used to visually select the most important reversals in a complex strain recording.

one phase to another should occur when the crack has a length approximately equal to one-quarter to one-half the radius of the notch or hole from which it starts, or at most a length of 5 mm (0.2 in.). This is an educated guess based on the assumptions that at this stage the crack tip will be closer to a field of nominal stress than to the field of local stress concentration, and that the crack will then be much longer than the width of the plastic zone ahead of it. The actual optimum transition depends on many factors that have not yet been investigated.

For the early stages computer programs have been published [1, 17] that calculate damage either for each reversal or for each closed loop or pair of ranges. Calculation of crack growth to failure is the next stage of the more sophisticated calculations, and this involves calculating the crack length just before fracture and the growth of cracks from the initial to that final length. The former calculation is relatively easy and not very critical. The final stage of the growth of cracks is so fast that a large difference in computed final crack length makes only a small difference in the computed life to fracture as is shown earlier in Section 5.5.4.

The calculation of crack growth from initial to final crack size is the critical part of the computation. It would be straightforward if one could use the crack growth data from tests with uniform range of load or of stress intensity factor, but this may not be accurate enough to justify the use of sophisticated methods. Crack growth depends not only on the range of stress intensity factor, but very significantly on the previous history, which may have left compressive or tensile stress fields. If compressive, the stress fields retard or arrest crack growth; if tensile they accelerate it.

The major interest in fatigue crack growth life predictions has been motivated by the fail-safe design philosophy of aircraft in which proper inspection periods must be predetermined. Current quantitative fatigue crack growth life predictions under variable amplitude conditions appear to be based on primarily three different models formulated in the early 1970s by Wheeler [22], Willenborg et al. [23], and Elber [24]. These models and their extensions or modifications require substantial numerical calculations with high speed computers and can be quite expensive. A fair knowledge of the load spectrum is needed to properly include interaction effects. The accuracy of the models has ranged from extremely good to extremely poor. In plain talk, however, the variable amplitude crack growth prediction models have not been developed to a point where simulated or actual field tests can be eliminated. We strongly reiterate the idea that the backbone of safe fatigue design is testing and field service monitoring and/or carefully analyzed experience. Prototype design and inspection period determination, however, have been and can be substantially enhanced through numerical life prediction calculations. A description of the three basic models follows. Substantial literature exists where these models have been analyzed, modified, expanded, and used in detail. Their use has been most prevalent in the aerospace field.

Wheeler Model [22].   The Wheeler model uses a modification of the basic functional relationship between constant amplitude crack growth rate and the stress intensity factor range

$$\frac{da}{dN} = f(\Delta K) \tag{10.2}$$

The Paris equation (Eq. 5.17) or the Forman equation (Eq. 5.21) could represent this functional relationship. The model also modifies the following basic summation, which implies the crack length after n load applications is

$$a_n = a_0 + \sum_{i=1}^{n} \Delta a_i = a_0 + \sum_{i=1}^{n} f(\Delta K_i) \tag{10.3}$$

where a_0 is the original crack length. The equation implies that sequence effects are omitted. The Wheeler modification introduces an empirical retardation parameter C_i that allows for retardation by reducing the crack growth rate following a high tensile load application where

$$C_i = \left[\frac{r_{yi}}{(a_{0L} + r_{0L}) - a_i} \right]^m \tag{10.4}$$

and r_{yi} = the plastic zone at any instant
   a_{0L} = the crack length at a high tensile load application
   r_{0L} = the plastic zone caused by the high load
   a_i = the crack length at any instant after the high load
   m = an empirical shaping exponent

The terms are defined schematically in Fig. 10.13. Equation 10.4 is the proper retardation parameter as long as

$$a_i + r_{yi} < a_{OL} + r_{OL} \tag{10.5}$$

If

$$a_i + r_{yi} \geq a_{OL} + r_{OL} \tag{10.6}$$

then

$$C_i = 1 \tag{10.7}$$

This implies that as long as the instantaneous plastic zone is within the high load affected plastic zone, crack growth retardation will be present. When the instantaneous plastic zone just reaches or passes beyond the high load affected plastic zone then retardation effects disappear. The resultant Wheeler crack length summation equation is

$$a_n = a_0 + \sum_{i=1}^{n} C_i f(\Delta K_i) \tag{10.8}$$

Values for r_{yi} for plane stress or plane strain from Eqs. 4.9 and 4.10 are

$$r_{yi} = \frac{1}{2\pi}\left(\frac{K_{\max i}}{S_y}\right)^2 \quad \text{or} \quad \frac{1}{6\pi}\left(\frac{K_{\max i}}{S_y}\right)^2 \tag{10.9}$$

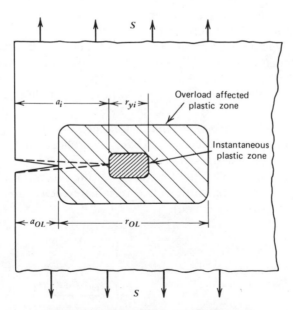

FIGURE 10.13   Crack tip plastic zones.

Values of r_{0L} for plane stress or plane strain from Eqs. 4.9 and 4.10 are

$$r_{0L} = \frac{1}{2\pi}\left(\frac{K_{0L}}{S_y}\right)^2 \quad \text{or} \quad \frac{1}{6\pi}\left(\frac{K_{0L}}{S_y}\right)^2 \quad\quad (10.10)$$

The exponent m shapes the retardation parameter C_i to correlate with test data. This value is obtained experimentally and can pertain to an entire known spectrum or can be varied to consider specific load ratios between a high load and a low load. Wheeler found values for m of 1.3 for D6AC steel and 3.4 for titanium in his experiments. If $m = 0$ then no retardation influence exists. The high speed computer numerical computation scheme requires crack growth summation one load application at a time.

Willenborg Model [23, 25]. The Willenborg model handles crack growth retardation by using an effective stress concept to reduce the applied stresses and hence the crack tip stress intensity factor. The model does not rely on empirically derived parameters except the constant amplitude crack growth parameters for a given material such as the Paris equation (Eq. 5.17) or Forman equation (Eq. 5.21). An effective value of the stress intensity factor range is computed by assuming some crack tip compressive self-stress is present after a high load. The effective ΔK is used in conjunction with the constant amplitude crack growth rate data and the CRACKS II [26] or EFFGRO [27] computer program to calculate crack growth life. The model assumes retardation decays over a length equal to the plastic zone r_{0L} from an overload as shown in Fig. 10.13. Any load greater than the preceding overload creates a new retardation condition independent of all preceding conditions. Only tensile loads are counted and compression loads are neglected. The model operates in terms of total length rather than crack length increments; hence the expression for the overload affected zone, a_p, becomes

$$a_p = a_{0L} + r_{0L} \quad\quad (10.11)$$

Retardation decays to zero when the instantaneous crack length a_i plus its associated yield zone r_{yi} exceeds a_p. Using this relationship, an expression for the decay in the self-stress is determined as follows. For any instantaneous crack length a_i following an overload, the stress, S_{req}, required to produce a yield zone, r_{req}, sufficient to terminate retardation is calculated as follows (plane stress assumed):

$$a_p = a_i + r_{req} = a_i + \frac{1}{2\pi}\left(\frac{K_{req}}{S_y}\right)^2 \qu\quad (10.12)$$

$$= a_i + \frac{1}{2\pi}\left(\frac{S_{req}\sqrt{\pi a_i}\,\alpha}{S_y}\right)^2 \qu\quad (10.13)$$

Solving for S_{req} gives

$$S_{req} = \frac{2S_y}{\alpha} \sqrt{\frac{a_p - a_i}{a_i}} \qquad (10.14)$$

The model postulates that the compressive self-stress due to the overload is the difference between the maximum stress, $S_{max\,i}$, occurring at a_i and the corresponding value of S_{req}

$$S_{self} = S_{req} - S_{max\,i} \qquad (10.15)$$

S_{self} depends on S_y, $\alpha\,a_i$, a_{OL}, r_{OL}, and $S_{max\,i}$. Once S_{self} has been determined, both $S_{max\,i}$, and $S_{min\,i}$ are reduced by this amount such that

$$S_{max\,eff} = S_{max\,i} - S_{self} = 2S_{max\,i} - S_{req} \qquad (10.16)$$

$$S_{min\,eff} = S_{min\,i} - S_{self}$$

$$= S_{min\,i} + S_{max\,i} - S_{req} \qquad (10.17)$$

where both $S_{max\,eff}$ and $S_{min\,eff}$ must be positive (tensile). If a negative value for $S_{max\,eff}$ or $S_{min\,eff}$ occurs, replace that term by zero. In terms of stress intensity factors

$$K_{max\,eff} = S_{max\,eff} \sqrt{\pi a_i \alpha} \qquad (10.18)$$

$$K_{min\,eff} = S_{min\,eff} \sqrt{\pi a_i \alpha} \qquad (10.19)$$

The effective load ratio R_{eff} becomes

$$R_{eff} = \frac{K_{min\,eff}}{K_{max\,eff}} \qquad (10.20)$$

Both ΔK_{eff} and R_{eff} can be determined and then da/dN can be calculated from, say, the Forman equation (Eq. 5.21) based on effective values

$$\frac{da}{dN} = \frac{A(\Delta K_{eff})^n}{(1 - R_{eff})K_c - \Delta K_{eff}} \qquad (10.21)$$

The above model can be handled range by range along with the CRACKS II computer program.

Elber Crack Closure Model [24].   The crack closure model is an empirically based model that uses an effective stress range concept to incorporate interaction effects in variable amplitude fatigue crack growth life predictions. Elber found that fatigue cracks subjected to plane stress tension-tension loading close before the remotely applied stress, S, becomes zero. Significant compressive self-stresses are transmitted across the crack at zero load. Elber assumed that crack extension occurs only when the applied stress is greater than the crack opening stress. Thus the controlling stresses in the crack growth process should be the maximum stress and the crack opening stress within a cycle. Elber found that the stress at which crack closure occurs was slightly different from that for crack opening. This difference is frequently neglected, and crack closure and opening stresses are often used interchangably. He attributed the crack closure to a zone of residual tensile deformations left behind the moving crack tip that are interrelated with crack tip compressive self-stresses.

The stress range that contributes to crack extension is called the effective stress range ΔS_{eff}, where

$$\Delta S_{eff} = S_{max} - S_{0p} \tag{10.22}$$

where S_{0p} is the crack tip opening stress determined experimentally. Defining a closure factor C_i as

$$C_i = \frac{S_{0p}}{S_{max}} \tag{10.23}$$

results in

$$\Delta S_{eff} = S_{max}(1 - C_i) \tag{10.24}$$

Using, for example, the Paris equation (Eq. 5.17) and replacing ΔK with ΔK_{eff} gives

$$\frac{da}{dN} = A(\Delta K_{eff})^n \tag{10.25}$$

$$= A(\Delta S_{eff} \sqrt{\pi a}\, \alpha)^n \tag{10.26}$$

$$= A[S_{max}(1 - C_i)\sqrt{\pi a}\, \alpha]^n \tag{10.27}$$

Equation 10.27 can then be used in a high speed computer numerical integration scheme to obtain cycle by cycle fatigue crack growth from some initial crack size to a final crack size. C_i is dependent on many factors, including material, thickness, temperature, corrosive environment, and stress peaks. Compressive stresses are assumed to be nondamaging since the crack tip is assumed to be closed during compression. Socie [28], however, has found crack opening stresses in compression for the SAE transmission history of Fig. 10.1b.

10.6 SIMULATING REAL HISTORIES IN THE LABORATORY

According to a proverb, experience is the best teacher. Unfortunately it is also a very expensive teacher. Much fatigue research has aimed at providing data and rules that distill the essence of experience so that the designer can calculate the sizes, shapes, and processing of his or her part and confidently expect them to last long enough without being too heavy, bulky, or expensive. For constant amplitude loading of well-known materials in well-known environments this can be done. For irregular loading we still lack tools for such confident analysis and prediction. To avoid the high cost of learning from field failures many companies and agencies have developed laboratory test methods that can approximate the results of field experience for irregular load histories. Six different methods are discussed below. They differ in philosophy and in cost. With sufficient experience in comparing laboratory data to field data they all can perform very well. They all require knowledge of the expected history as primary input. Some agencies, such as the U.S. Air Force, provide a spectrum or history of use. For other customers, like drivers of cars, it is very difficult to decide on a "representative" load history.

In order of increasing physical complexity the six methods discussed are: (1) characteristic constant amplitude, (2) block testing, (3) condensed histories, (4) truncated histories, (5) complete histories, and (6) statistically simulated histories.

10.6.1 Testing with a Characteristic Constant Amplitude

A characteristic constant amplitude, based on experience, serves very well if it can be validated by field data. For automobile suspension springs, for instance, constant amplitude tests to a few hundred thousand cycles of maximum possible deflection reproduce the field history well enough. The reasons for this are: (1) the large amplitudes actually do most of the fatigue damage and (2) field use is so diverse that any one field history would be just as wrong as this constant amplitude test.

An "equivalent" constant amplitude has been defined as that constant amplitude that produces the same median life, in numbers of reversals, as the real history. Because it is difficult to know the median life for the real history and because important large ranges will be omitted to obtain equal numbers of reversals we see little merit in an "equivalent" constant amplitude test.

10.6.2 Block Testing

The real history is replaced by a number of "blocks" of constant maximum and minimum load or deflection as shown in Figs. 5.2d and 5.2e. The series of blocks form a program. In principle, the program contains the same number of reversals as the history, and its blocks approximate the distribution of peaks and valleys. In practice large numbers of small ranges, well below the fatigue limit, are omitted. Six to 10 blocks provide adequate approximations. The sequence of blocks is important. A random or pseudorandom sequence will minimize undesirable sequence effects. High amplitude ranges that occur only seldom are added once in every nth program. It has been found that a test should contain at least a dozen repetitions of the program in order to represent a real history. This method of testing has been widely used in the aircraft industry.

10.6.3   Testing with Condensed Histories

Modern electronic–hydraulic controls are capable of producing practically any prescribed sequence of loads, deflection, or strains. They are expensive and must be operated at relatively low frequencies if large deflections must be included in the program. Condensed histories include selected peaks and valleys of the real history in their real sequence and omit many peaks and valleys. It has been shown that with only 2 percent of the reversals, properly selected, the fatigue life to the appearance of 2.5 mm deep cracks was practically the same as with all the reversals for histories shown in Fig. 10.1 [12, 29]. The selection of the most significant peaks and valleys is done by racetrack counting or by editing a rainflow count to retain only the largest ranges.

10.6.4   Truncated Histories

Histories can be truncated by omitting all ranges smaller than a given amount. If this is done after rainflow counting it amounts to the same as condensed histories. If it is done without regard to overall ranges it may lead to serious errors because a small range (say, 10 percent of the maximum range) may easily add 5 percent to the highest peak or lowest valley and thus increase the damage 30 percent or more. The same magnitude, discarded by racetrack or rainflow counting, might have contributed only one-millionth (0.0001 percent) of the damage of the maximum range. (Above numbers assumed an S-N slope of $-\frac{1}{6}$, or damage proportional to the sixth power of amplitudes.)

10.6.5   Complete Histories

Suitable test machines can take a record of a load history on tape and play it back to the test specimen over and over again. This produces a good test, but it may be unnecessarily expensive in time in view of the capabilities of condensed histories.

10.6.6   Statistically Simulated Histories

Repeating a recorded history over and over again may not be the best way to test parts. It may, for instance, omit or exaggerate some sequence effects. To overcome this one can arrange truly random input to the test machine with prescribed parameters, such as the distribution of peaks and valleys or of amplitudes and means. Other parameters that might be prescribed are

RMS value, maximum value, and harmonic component spectrum, for instance, ''white Gaussian noise.''

10.7 DOS AND DON'TS IN DESIGN

1. Do learn as much as possible from service failures. They may validate or invalidate your assumptions about service loads.
2. Do ask yourself whether sequence effects are likely to be important.
3. Do determine (by test or by agreement) the service load histories for which you are designing.
4. Do allow margins for error in keeping with the certainty or uncertainty of your assumptions.
5. Do place prototypes or early production machines in severe service and follow them very carefully to obtain load histories and to detect weak spots.

REFERENCES FOR CHAPTER 10

1. R. M. Wetzel, Ed., *Fatigue under Complex Loading,* SAE, 1977.
2. L. E. Tucker and S. L. Bussa, ''The SAE Cumulative Fatigue Damage Program,'' *Fatigue under Complex Loading,* SAE, 1977, p. 1.
3. A. Palmgren, ''Durability of Ball Bearings,'' *ZDVDI,* Vol. 68, No. 14, p. 339 (in German).
4. M. A. Miner, ''Cumulative Damage in Fatigue,'' *Trans. ASME, J. Appl. Mech.,* Vol. 67, Sept. 1945, p. A159.
5. T. L. Salt, ''A Designers Approach to the Fatigue Failure Mechanism,'' *Ann. Reliability Maintainability Symp.,* 1973, p. 437.
6. T. L. Salt, U.S. Patents 3,887,987 and 3,908,447 (1975).
7. T. H. Topper, B. I. Sandor, and J. Morrow, ''Cumulative Fatigue Damage under Cyclic Strain Control,'' *J. Mater.,* Vol. 4, No. 1, March 1969, p. 200.
8. J. H. Crews, Jr., ''Crack Initiation at Stress Concentrations as Influenced by Prior Local Plasticity,'' *Achievement of High Fatigue Resistance in Metals and Alloys,* ASTM STP 467, 1970, p. 37.
9. R. I. Stephens, D. K. Chen, and B. W. Hom, ''Fatigue Crack Growth with Negative Stress Ratio Following Single Overloads in 2024-T3 and 7075-T6 Aluminum Alloys,'' *Fatigue Crack Growth under Spectrum Loads,* ASTM STP 595, 1976, p. 27.
10. D. V. Nelson and H. O. Fuchs, ''Predictions of Cumulative Fatigue Damage Using Condensed Load Histories,'' *Fatigue under Complex Loading,* SAE, 1977, p. 163.
11. H. O. Fuchs, ''Discussion: Nominal Stress or Local Strain Approaches to Cumulative Damage,'' *Fatigue under Complex Loading,* SAE, 1977, p. 203.

12. H. O. Fuchs, D. V. Nelson, M. A. Burke, and T. L. Toomay, "Shortcuts in Cumulative Damage Analysis," *Fatigue under Complex Loading,* SAE, 1977, p. 145.

13. A. Teichmann, "The Strain Range Counter," Vickers-Armstrong Aircraft Ltd., Tech. Office VTO/M/46, April 1955.

14. J. E. Hayes, "Fatigue Analysis and Fail-Safe Design," *Analysis and Design of Flight Vehicle Structures,* E. F. Bruhn, Ed., Tri-State Offset Co., Cincinnati, OH, 1965, p. C 13-1.

15. N. E. Dowling, W. R. Brose, and W. K. Wilson, "Notched Member Fatigue Life Predictions by the Local Strain Approach," *Fatigue under Complex Loading,* SAE, 1977, p. 55.

16. S. Downing, D. Galliart, and T. Berenyi, "A Neuber's Rule Fatigue Analysis Procedure for Use with a Mobile Computer," *Fatigue under Complex Loading,* SAE, 1977, p. 189.

17. D. F. Socie, "Fatigue Life Prediction Using Local Stress–Strain Concepts," *Exp. Mech.,* Vol. 17, No. 2, Feb. 1977, p. 50.

18. M. Matsuishi and T. Endo, "Fatigue of Metals Subjected to Varying Stress," paper presented to Japan Society of Mechanical Engineers, Fukuoka, Japan, March 1968.

19. S. L. Bussa, N. J. Sheth, and S. R. Swanson, "Development of a Random Life Prediction Model," Ford Engineering Technology Office Report No. 70-2204, May, 1970.

20. J. M. Barsom, "Fatigue Crack Growth under Variable-Amplitude Loading in Various Bridge Steels," *Fatigue Crack Growth under Spectrum Loads,* ASTM STP 595, 1976 p. 217.

21. J. M. Potter, "Spectrum Fatigue Life Predictions for Typical Automotive Load Histories," *Fatigue under Complex Loading,* SAE, 1977, p. 107.

22. O. E. Wheeler, "Spectrum Loading and Crack Growth," *Trans., ASME, J. Basic Eng.,* Vol. 94, 1972, p. 181.

23. J. Willenborg, R. M. Engle, and H. A. Wood, "A Crack Growth Retardation Model Using an Effective Stress Concept," WPAFB, TM-71-1-FBR, 1971.

24. W. Elber, "The Significance of Fatigue Crack Closure," *Damage Tolerance in Aircraft Structures,* ASTM STP 486, 1971, p. 230.

25. R. M. Engle and J. L. Rudd, "Analysis of Crack Propagation under Variable-Amplitude Loading Using the Willenborg Retardation Model," *Proc. AIAA/ASME/SAE 15th Structures, Structural Dynamics and Materials Conference,* Las Vegas, Nevada, April, 1974.

26. R. M. Engle, "CRACKS II Users Manual," AFFDL TM-173-FBE, Aug. 1974.

27. J. B. Chang, "Improved Methods for Predicting Spectrum Loading Effects," Rockwell International, Los Angeles Division, first quarterly report, NA-78-491, May 1978.

28. D. F. Socie, "Estimating Fatigue Crack Initiation and Propagation Lives in Notched Plates under Variable Loading Histories," Ph.D. Thesis, University of Illinois, 1977.

29. D. F. Socie and P. Artwahl, "Effects of Spectrum Editing on Fatigue Crack

Initiation and Propagation in a Notched Member," University of Illinois, College of Engineering Report FCP No. 31, Dec. 1978.

PROBLEMS FOR CHAPTER 10

1. A smooth machined uniaxial hot-rolled 1020 steel rod is subjected to five fully reversed blocks of nominal stress cycling as indicated below. The blocks are then repeated. Predict the expected fatigue life of the part.

S_a (MPa)	350	300	250	200	125
Applied cycles	5	50	500	5000	10,000

   If the part had a circular groove with $K_t = 1.5$ and a radius of 5 mm predict the expected fatigue life.
2. Repeat problem 1 for the following:

   (a) RQC-100 steel.
   (b) 4340 steel with $S_u = 1250$ MPa.
   (c) 4340 steel with $S_u = 1470$ MPa.
   (d) 2024-T3 aluminum.
   (e) 7075-T6 aluminum.

3. Assume the ordinate in Fig. 10.1d is twice the nominal stress in MPa (2:1 scale). Construct a peak and valley histogram of occurrences versus stress. Choose a small stress increment for the absissa because of the small number of reversals. Then construct a cumulative distribution of peaks and valleys. Can you do the above for stress ranges?
4. Rainflow count the history of Fig. 10.1d and plot the range histogram of occurrences versus stress. Then construct a cumulative distribution of ranges.
5. How many blocks of Fig. 10.1d loading can be applied to a small smooth uniaxial rod of RQC-100 steel if the given axis has a scale of 200 units = 100 MPa? Repeat the calculations if a notch of $K_t = 1.5$ and radius equal to 5 mm had existed.
6. Repeat problem 5 using the following:

   (a) 9262 steel with $S_u = 1000$ MPa.
   (b) 4340 steel with $S_u = 1250$ MPa.
   (c) 4340 steel with $S_u = 1470$ MPa.

7. A typical ground–air–ground (GAG) spectrum is shown in Fig. 5.2a. Substantial comparative testing of several prototype solutions is to be done. Simplify the given spectrum in order to make the tests. Provide five optional simplified spectra with the advantages and limitations of each. Do any of your five spectra have some commonality?

8. In problem 16, Chapter 5 reverse the order of loading and then solve parts *a* to *d* without considering interaction effects.

9. Repeat problem 8, but use retardation influence.

10. How many blocks of Fig. 10.1*d* loading can be applied to a wide center cracked panel with an initial crack length $2a = 2$ mm? The given axis has a scale of 200 units = 50 MPa. Use the following materials:

   (*a*) Hot-rolled 1020 steel.
   (*b*) 4340 steel with $S_u = 1250$ MPa.
   (*c*) 2024-T3 aluminum.
   (*d*) 7075-T6 aluminum.

11. Determine when and explain why a tensile overload such as that in Fig. 10.4*a* can (*a*) be beneficial, (*b*), be detrimental, and (*c*) have a negligible effect on fatigue life of parts.

12. Same as problem 11, but consider the overloads of Figs. 10.4*b* and 10.4*c*.

CHAPTER **11**

ENVIRONMENTAL EFFECTS

Environmental effects on fatigue of metals may be more severe than sharp stress concentrations or almost harmless. Quantitative fatigue life predictions are often not possible because of the many interacting factors that influence environmental fatigue behavior and lack of significant data. For example, in corrosion fatigue, frequency effects can be quite substantial, while in noncorrosive environments frequency is usually a second order concern. Design against fretting fatigue, particularly in mechanical joints, has not been quantified. At elevated temperatures mean stress effects are extremely complex because of interaction among creep, fatigue, and environment. The linear elastic stress intensity factor K also has more limitations at elevated temperatures because of appreciable plasticity. Substantial reduction in fracture toughness can occur at low temperatures, which reduces critical crack sizes at fracture. Irradiation can reduce both fatigue resistance and fracture toughness. Despite these large difficulties fatigue design with environmental considerations must be accomplished. This places great emphasis on real-life product testing, inspection, service history analysis, experience, and accelerated acquisition of experience. The above environmental topics of corrosion, fretting, temperature, and irradiation effects on fatigue behavior are covered in this chapter. These topics are of appreciable importance to many engineering fields, including aerospace, naval, ground vehicles, and nuclear and mechanical systems.

11.1 CORROSION FATIGUE

There are many environments that affect fatigue behavior; however, most engineering components and structures interact with air, water, and salt water. Thus most of our following discussion involves these three environments. The principles, however, are applicable to other corrosive environments.

Corrosion fatigue refers to the joint interaction of corrosive environment and repeated stressing. The combination of both acting together is more detrimental than either acting separately. That is, repeated stressing accelerates the corrosive action and the corrosive action accelerates the mechanical fatigue mechanisms. Under static loads, corrosive environments may also be detrimental, particularly in higher strength alloys. Environmental assisted fracture under static loading is called "stress corrosion cracking." We briefly consider this topic before covering corrosion fatigue.

11.1.1 Stress Corrosion Cracking

Let us consider a number of identical precracked test specimens (for example Figs. 5.6 g to 5.6j) or components, made from the same medium or high strength alloy. Subject each specimen to a different initial force that causes a different initial plane strain stress intensity factor, K_{Ii}. Keep those initial forces applied to each specimen until fracture or test termination. Simultaneously expose the specimens or components to a corrosive environment. Depending on the initial stress intensity factor, the cracks will grow at different rates. The time to fracture can be monitored, and results similar to these shown in Fig. 11.1 are obtained. At very short life the initial stress intensity factor is essentially the same as the fracture toughness K_{Ic} obtained without the corrosive environment. As K_{Ii} is reduced, the time to fracture increases. A limiting threshold value, K_{ISCC}, is finally obtained, which is the stress intensity factor below which a crack is not observed to grow under the specific environment. The subscripts refer to mode I or plane strain conditions and stress corrosion cracking. The value 10^3 hours has often been chosen as a minimum termination time for K_{ISCC} using constant load tests. With constant crack opening displacement tests the time of testing can be shorter. K_{ISCC} has become a common design property and substantial K_{ISCC} data have been tabulated in the *Damage Tolerant*

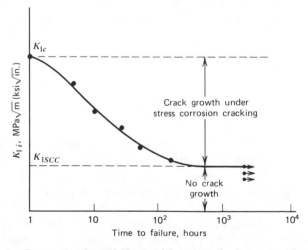

FIGURE 11.1 Determination of K_{ISCC} with precracked constant load specimens.

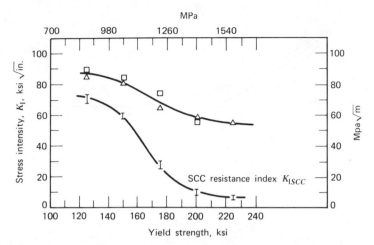

FIGURE 11.2 Effect of yield strength on K_{ISCC} and fracture toughness in 4340 steel [2] (reprinted with permission of the National Association of Corrosion Engineers). A1S1 4340 tested in flowing seawater (Key West). Fracture orientation: WR, Fracture toughness index: ($\square$) K_{IX} (dry brake), ($\triangle$) $K_{I\delta}$.

Handbook [1]. Values of K_{ISCC} range from approximately 0.1 to 1.0 times K_{Ic}.

Figure 11.2 shows typical values of K_{ISCC} as a function of yield strength for 4340 steel immersed in a salt water environment [2]. As the yield strength increased, K_{ISCC} decreased even more rapidly than an approximate fracture toughness index labeled K_{Ix} or $K_{I\delta}$. Thus, in general, as the yield strength increases, the susceptibility to stress corrosion cracking becomes more pronounced. Lower ratios of K_{ISCC}/K_{Ic} usually occur with the higher yield strength materials.

The above discussion on stress corrosion cracking is very much related to corrosion fatigue. It implies that repeated loads are not needed for cracks to extend if applied stress intensity factors are above K_{ISCC}. Thus a complex interaction exists between static and repeated loads in the presence of corrosive environments. Corrosion fatigue cracks can grow at stress intensity factors below K_{ISCC}.

11.1.2 Stress–Life Behavior, S–N

Since most fatigue data are obtained in laboratory air, we have primarily only corrosion fatigue data at our disposal because the air environment can certainly be corrosive with relative humidity or moisture content and oxygen the two major influencing factors. However, laboratory air fatigue data are often used as a reference for comparing other environmental effects on fatigue.

Figure 11.3 schematically shows typical constant amplitude S–N diagrams

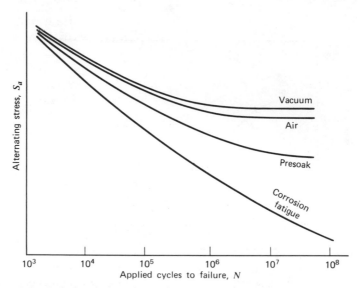

FIGURE 11.3   Relative fatigue behavior under various environmental conditions.

obtained at room temperature under four key room temperature conditions. The relative fatigue behavior shown is very realistic. If air tests are taken as the reference, it is seen that a vacuum can have a small beneficial effect primarily at long fatigue life. This benefit, however, depends on the material and the reference air environment. High humidity air can be quite detrimental to fatigue behavior. This is particularly true with aluminum alloys. Presoaking fatigue specimens in a corrosive liquid followed by testing in air often causes a detrimental effect. However, the combination of simultaneous environment and repeated stressing causes the most drastic decrease in long-life fatigue strengths as shown in Fig. 11.3. The relatively flat long-life S–N behavior that can occur in air or vacuum is eliminated under corrosion fatigue conditions. The long-life corrosion fatigue strengths in various environments can vary from approximately 10 to 100 percent of air fatigue strengths. For some materials, air fatigue properties can even be improved by some environments as a result of surface protection from oxygen and water vapor. At short lives, all four tests conditions shown in Fig. 11.3 tend to converge. This is primarily due to insufficient time for corrosion to be effective.

The mechanisms of corrosion fatigue are not well understood. It is an electrochemical process dependent on the environment/material/stressing interaction. Pits which act as stress raisers for cracks to initiate more readily, can form very early in life. A greater number of surface fatigue cracks usually occur under corrosive fatigue conditions and these tend to grow across grains. As the cracks get larger they also tend to grow in mode I perpendicular to the maximum tensile stress. Fracture surfaces are often discolored in the fatigue crack growth region. Very little loss of material due to corrosion occurs and thus loss of material is not a major contributing factor. Both crack initiation life and crack growth life are reduced under corrosion fatigue conditions. Frequency effects are important and alter fatigue behavior as a result of the time dependent nature of corrosion fatigue. Only at very short lives or high crack growth rates are frequency effects small. However, under these conditions air and corrosion fatigue behavior in general are similar. Corrosion fatigue also depends on how the corrosive environment is applied. For example, specimens submerged in fresh or salt water have better corrosion fatigue resistance than those subjected to a water spray, drip, or wick or those submerged in continuously aereated water. This is due to the great importance of oxygen and the formation of oxide films in the corrosion fatigue process.

The above description only partially explains the results in Fig. 11.3. The vacuum is beneficial because of the elimination of water vapor and oxygen. The presoaking can cause pits to form, which then act as stress raisers. Thus the presoaking can cause fatigue behavior similar to that for notched specimens. The low corrosion fatigue curve is the result of the interaction among environment, material, and stressing. Protective oxide films that can form at pits and crack tips under only corrosive conditions can be continually broken by repeated stresses such that new fresh surfaces are continually exposed to the corrosive environment. Deep cracks in large thick components may not be exposed to this environment interaction and may grow in a manner similar to the growth of cracks in air environment.

McAdam [3] in the late 1920s carried out a comprehensive air and corrosion fatigue program with carbon, low alloy, and chromium steels using rotating bending specimens subjected to fresh water spray. The results of these tests are summarized in Fig. 11.4 [4], where fatigue strengths in air and water at 2×10^7 cycles are plotted against ultimate tensile strength.

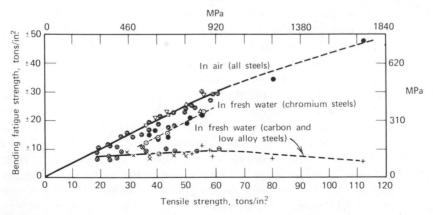

FIGURE 11.4 Influence of ultimate tensile strength on fresh water corrosion fatigue limits [4] (reprinted with permission of Pergamon Press Ltd.).

The results show the trend of unnotched fatigue strengths in air to increase with ultimate strength. The fatigue strengths in water for the carbon and low alloy steels, however, are almost independent of the ultimate strength. These quantitative data substantiate the qualitative schematic behavior shown in Fig. 5.9. All the corrosion fatigue strengths for these annealed and quenched and tempered carbon and low alloy steels were between 85 and 210 MPa (12 and 30 ksi), yet their ultimate tensile strengths varied from 275 to 1720 MPa (40 to 250 ksi). These very low corrosion fatigue strengths are one of the most often overlooked aspects in fatigue design. The corrosion fatigue strengths for steels containing 5 percent or more chromium were much better, but they were still less than the air fatigue strengths. The above results were obtained at 24 Hz, which is very high. At lower frequencies, where more time is available, these values could be even lower.

Table A.5 includes corrosion fatigue strengths for selected engineering alloys subjected to water or salt environments [4]. These data only serve to illustrate corrosion fatigue effects and should not be used as design values. In general, salt water is more detrimental than fresh water, but variations in water chemistry can affect fatigue behavior. Water and salt water were detrimental to fatigue strengths in all alloys listed except for copper and phosphor bronze. Detrimental effects are shown for steels, brass, aluminum, magnesium, and nickel. Additional data are available in reference 4. In general, those materials with the best corrosion resistance in a specific environment also had the best corrosion fatigue resistance. Thus a key to determining good fatigue design is to look for high corrosion resistance.

11.1.3   Crack Growth Behavior, da/dN–ΔK

Fatigue crack growth plays an important role in corrosion fatigue; however, insufficient data and understanding exist. The lack of realistic crack growth data is due to the great expense and time needed to obtain low crack growth rates. Normal frequencies of 5 to 25 Hz, which are often used in noncorrosive environments, are too high for realistic corrosion fatigue crack growth behavior. Frequencies of 1 to 10 cpm (0.017 to 0.17 Hz) are required to obtain realistic corrosion fatigue crack growth rates. This implies months of test time are needed that usually have not been taken. Thus most corrosion fatigue crack growth rates available in the literature are for greater than 2.5×10^{-8} m/cycle (10^{-6} in./cycle), which comprises regions II and III of the sigmoidal curve of Fig. 5.19. Threshold levels ΔK_{th} for corrosion fatigue crack growth are usually unknown. At very high crack growth rates, where less time is available for the interaction between corrosion and repeated stresses, the differences between corrosion and noncorrosion rates are often small. References 5 and 6 provide comprehensive data on corrosion fatigue crack growth behavior.

The detrimental effect of moisture on air corrosion fatigue of aluminum alloys is indicated in Section 11.1.2. Figure 11.5, from Bowles [7], shows just how important moisture is on fatigue crack growth of 2024-T3 and 7075-T6 aluminum sheet. Under both laboratory air and dry air conditions, the 2024-T3 alloy has better crack growth properties than 7075-T6, but the differences are much greater under dry air conditions. An order of magnitude difference between laboratory air and dry air da/dN occurs at lower ΔK values for 2024-T3 aluminum, while only a factor of 2 or less is found for the 7075-T6. These differences are quite significant considering the very high frequency of 20 Hz used in these tests. At higher crack growth rates the laboratory air and dry air curves for each alloy tend to converge. With this large difference from just moisture alone, it becomes quite evident how complex corrosion fatigue can be.

The effect of fresh natural flowing seawater and electrochemical potential

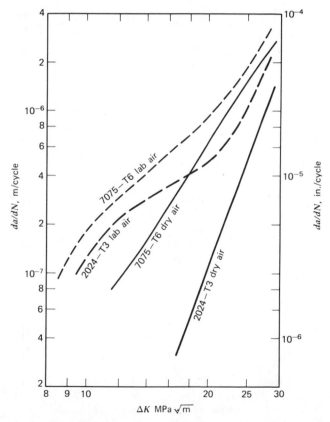

FIGURE 11.5 Crack growth rates in dry and laboratory air for 2024-T3 and 7075-T6 aluminum, 20 Hz [7] (reprinted with permission of the Department of Aerospace Engineering, Delft University).

on fatigue crack growth of a common marine alloy HY-130 steel obtained by Crooker et al. is shown in Fig. 11.6 [8]. The three seawater curves were obtained at 1 cpm (0.017 Hz) or 10 cpm (0.17 Hz), while the reference ambient air data were obtained at 0.5 to 5 Hz. In all three seawater test conditions the fatigue crack growth rates were greater than in ambient air. Higher negative electrochemical potential and lower frequency significantly increased crack growth rates. The effect of frequency obtained by Vosikovsky [9] is shown in Fig. 11.7 for X-65 line-pipe steel subjected to 3.5 percent salt water. He also obtained higher crack growth rates at lower frequencies, but only at the higher ΔK values. The reference air data scatter is shown by the dashed lines.

Imhof and Barsom [10] found that increasing the yield strength in 4340 steel increased corrosion fatigue crack growth rates in 3 percent NaCl as

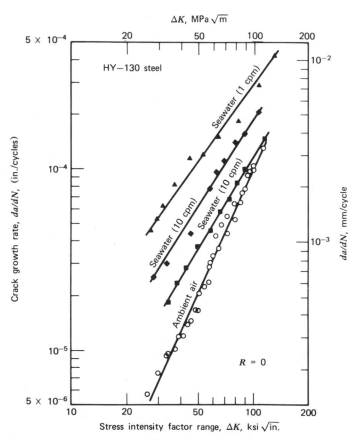

FIGURE 11.6 Corrosion fatigue crack growth rates for HY-130 steel in natural flowing seawater [8] (reprinted by permission of the American Society for Testing and Materials). (○) Ambient air (30 cpm), (■) seawater (-665 mV), (◆ seawater (-1050 mV), (▲) seawater (-1050 mV).

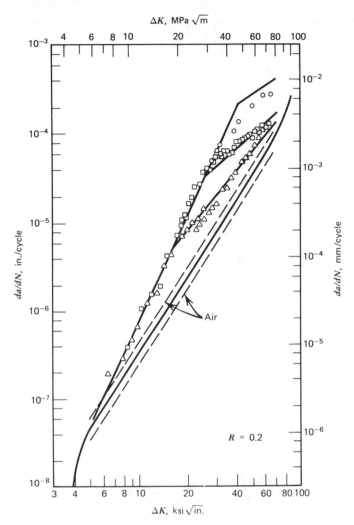

FIGURE 11.7 Corrosion fatigue crack growth rates for X-65 line-pipe in 3.5 percent salt water [9] (reprinted with permission of the American Society of Mechanical Engineers). ($\triangle$) 1 Hz, ($\square$) 0.1, ($\triangle$) 0.01.

shown in Fig. 11.8. Crack growth rates in air for the two steels were essentially the same. K_{ISCC} for the two steels are shown in Fig. 11.8 and all corrosion fatigue crack growth data were obtained at stress intensity levels below these respective values.

Mean stress effects on 3 percent NaCl corrosion fatigue crack growth behavior indicate that for $R \geq 0$, increasing the R ratio can cause an increase in fatigue crack growth rates. Chu and Macco [11] found the effect to be small in 5456-H117 aluminum, while Bucci [12] and Ryder et al. [13] found much greater effects in Ti–8Al–1Mo–IV and Ti–6Al–4V, respectively.

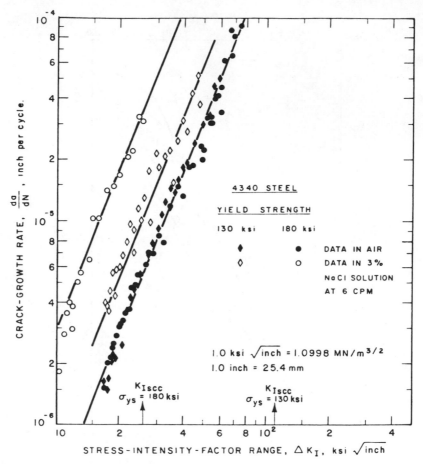

FIGURE 11.8 Corrosion fatigue crack growth rates for 4340 steel in 3 percent NaCl solution [10] (reprinted by permission of the American Society for Testing and Materials).

11.1.4 Protection against Corrosion Fatigue

The principal way we can reduce corrosion fatigue problems is to choose materials that resist corrosion in the expected environment. Increased corrosion fatigue resistance can also be accomplished through various surface treatments, such as shot-peening, cold working, and nitriding, which induce desirable surface compressive self-stresses. Anodic coatings have usually been beneficial, while cathodic coatings have been detrimental. Zinc and cadmium coatings are anodic to steels and have produced improved corrosion fatigue resistance. Zinc coatings have provided the better improvements. Chromium and nickel are cathodic to steels, and electrolytic plating with these metals produces undersirable tensile self-stresses, hairline surface cracks, and possibly hydrogen embrittlement.

Surface coatings such as paint, oil, polymers, and ceramics can protect against corrosive air and liquid environments if they remain continuous. However, under service conditions for many components and structures it is difficult or maybe impossible for these coatings to retain complete continuity. Broken or disrupted coatings can eliminate the beneficial effects.

Oxide coatings can be beneficial to corrosion fatigue resistance. Cladding of higher strength aluminum alloys with a pure aluminum surface layer (alcladding) has caused substantial increases in corrosion fatigue resistance of the base alloy. They often decrease air fatigue resistance, however. Shot-peening in conjunction with oxide coatings has caused even greater increases in corrosion fatigue resistance. In fact, shot-peening alone is quite beneficial to corrosion fatigue resistance, as well as to air fatigue resistance.

Corrosion inhibitors that form an adherent corrosion resistant chemical film on the metal surface have been somewhat successful. Chromates and dichromates have been widely used.

11.1.5 Dos and Don'ts in Design

1. Do consider that materials are susceptible to stress corrosion cracking under static loads when stress intensity values are greater than K_{ISCC}, which varies from about 0.1 to 1.0 times K_{Ic}.
2. Don't relate water or salt water corrosion fatigue resistance of steels to ultimate tensile strength. Many carbon and low alloy steels have similar corrosion fatigue strengths in water and salt water and thus high strength steels may not be advantageous unless surface compressive self-stresses and/or protective coatings are used. Long-life corrosion fatigue strengths in water and salt water can vary from about 5 to 40 percent of the ultimate strength.
3. Do obtain better corrosion fatigue resistance by choosing a material that exhibits low corrosion in the service environment.
4. Do consider stainless steels for better corrosion fatigue resistance; but their resistance is reduced relative to fatigue resistance in air.
5. Don't overlook the deleterious effects of humidity on fatigue resistance, particularly in aluminum alloys.
6. Do consider the many factors that can improve corrosion fatigue resistance, such as shot-peening, surface cold working, nitriding, anodic coatings cadmium and zinc, cladding, paint, oil, ceramic and polymeric coatings, and chemical inhibitors.

11.2 FRETTING FATIGUE

The nature of fretting-induced fatigue failures is not well understood and the terms used to describe the phenomena are not universal. Terms such as fretting, fretting corrosion, fretting fatigue, fretting corrosion fatigue, and fretting-initiated fatigue are commonplace. All terms given, however, include the word fretting and all terms involve the behavior of two surfaces in contact subjected to small repeated relative motion. We define these terms as follows and note that they are often interelated and interchanged.

Fretting: A surface wear phenomenon occurring between two contacting surfaces having oscillating relative motion of small amplitude.

Fretting corrosion: A form of fretting in which chemical reaction predominates.

Fretting corrosion fatigue: The combined action of fretting, chemical reaction, and fatigue.

Fretting fatigue: The combined action of fretting and fatigue, which can also include chemical reactions

Fretting-initiated fatigue: Can mean the same as fretting fatigue or can refer to fatigue at sites of previously formed fretting pits.

We primarily use the words fretting and fretting fatigue and let it be implied that corrosion products may also be involved, which is the usual case for metals operating in air.

Design against fretting fatigue is probably the least quantitative of all fatigue topics, yet it is involved with all assembled structures and components that receive or produce repeated motion. Examples include riveted, bolted, pinned, and lug fasteners, shrink and press-fits, splines, keyways, clamps, universal joints, bearing/shaft/housing interfaces, gear/shaft interfaces, oscillatory bearings, fittings, leaf springs, and wire rope. Fretting involves wear mechanisms and can occur with less than 10^{-5} mm of relative slip between mating surfaces. It can result in seizure of mating parts, loss of fit, or fatigue failures. Seizure is caused by the buildup of fretting debris that does not escape, while loss of fit is due to fretting debris escaping from the contacting surfaces. Fatigue failures occur from cracks initiating at the interface region and then propagating under cyclic stresses until fracture. The reduction in fatigue resistance due to fretting is of equal importance to notch effects and corrosion fatigue. Mann [14] indicates that fretting fatigue strengths may be as low as 5 to 10 percent of the base unnotched fatigue strengths, which implies fatigue strength reduction factors of between 10 and 20 may occur. Table 11.1 [15] shows that fatigue strength reduction factors for various combinations of metals in contact range from almost 1 to 5. This type of fatigue strength reduction certainly justifies the importance of considering fretting fatigue in design.

Fretting fatigue is of less importance in short-life components because of the number of repeated cycles necessary to form the fretting damage, initiate the cracks, and grow the cracks to fracture. That is, fretting fatigue is of greatest importance at long life of more than 10^4 or 10^5 cycles. The major factors affecting fretting action are the normal pressure between the mating

TABLE 11.1   Fatigue Strength Reduction Factors Produced by the Fretting of Various Materials against Steels and Aluminum Alloys [15][a]

Specimen	Hardness VHN	Fatigue strength (MPa)	Clamp	Fretting fatigue strength (MPa)	Strength reduction factor	Ref.
Carbon steels						
0.1C steel	137	172	0.1C steel	122	1.41	
			Brass	95	1.83	c
			Zinc	137	1.25	
0.33C steel	165	372	0.33C steel	254	1.48	d
0.4C steel	420	550	0.2C steel	450	1.23	
			0.4C steel	257	2.00	
			70/30 brass	325	1.70	e
			Al–4.4Cu–0.5Mn–1.5Mg	500	1.10	
0.7C steel, cold-drawn	365	525	0.7C steel, cold drawn	147	3.18	
0.7C steel, normalized	270	371	0.7C steel, normalized	178	2.08	f
Alloy steels						
0.25Cr–0.25Ni–1.0Mn	285	372	0.1C steel	294	1.62	
			18Cr–8Ni steel	264	1.80	c
1.3Cr–2.6Ni–0.4Mo	217	304	3 Si steel	241	1.19	g
1.1Cr–3.7Ni–0.4Mo	176	272	18Cr–8Ni steel	212	1.28	
			Aluminum	238	1.14	g
0.6Cr–2.5Ni–0.5Mo	330	542	0.6Cr–2.5Ni–0.5Mo	124	4.14	h
1.4Cr–4.0Ni–0.3Mo	510	850	1.4Cr–4.0Ni–0.3Mo	240	3.55	h
Aluminum alloys						
Al–Cu–Mg	—	276	Al–Cu–Mg	99	2.70	i
Al–4.4Cu–0.5Mn–1.5Mg	140	159	Al–4.4Cu–0.5Mn–1.5Mg	32	1.92	e
Al–4.4Cu–0.8Mn–0.7Mg	160	134	Al–4.4Cu–0.8Mn–0.7Mg	49.5[b]	2.72	
			Mild steel	35.6[b]	3.78	j
Al–4Cu	117	83.5	Al–4Cu	52.5	1.59	k
Copper alloys						
70/30 brass	175	139	70/30 brass	93	1.50	l

[a] (With permission of Pergamon Press).
[b] 193 MPa mean stress.
[c] Oding and Stepanov (1964).
[d] Cornelius and Bollenrath (1941).
[e] Corten (1955).
[f] Waterhouse and Taylor (1971).
[g] Oding and Ivanova (1956).
[h] Field and Waters (1967).
[i] Cornelius (1944).
[j] Fenner and Field (1958).
[k] Waterhouse (1972).
[l] Wharton, Taylor and Waterhouse

surfaces, the relative motion amplitude, friction, self-stresses, environment, mating materials, and number of applied cycles. Fretting damage has been compared both on the basis of material weight loss and by fatigue strength reduction. These two criteria can imply opposite effects, which can be confusing. We emphasize the reduction in fatigue resistance, since weight losses are generally small and not the major concern in fatigue design.

11.2.1 Mechanism of Fretting Fatigue

A fretting fatigue failure of a stainless steel strap subjected to alternating bending loads and rubbing against a cadmium plated member is shown in Fig. 11.9 [16]. The fretting effect was greater than the combined effect of the reduced cross section at the hole and its accompanying stress concentration factor. Substantial fretting debris (dark regions) are found in the rubbing area. These are primarily oxides of the base metal and for steels the iron oxide is rust colored, while for aluminum and magnesium alloys the fretting debris is black. Micro fatigue cracks initiate in the fretting zone often at an interface between fretted and nonfretted regions. Multiple microcracks can initiate both from shallow dishlike pits and from deeper pits that form under the fretting action. These microcracks are usually oblique to the surface, similar to the stage I cracks shown in Fig. 3.16. Eventually they become a single predominant mode I crack perpendicular to the surface or perpendicular to the maximum applied tensile stress. Once the crack tip leaves the vicinity of the fretting region it is no longer controlled by the fretting process but rather by the local stress field near the crack tip. LEFM principles, and hence the stress intensity factor range, ΔK, then play the predominant role. If ΔK is greater than threshold levels, the crack grows under cyclic loading. The stress intensity factor and the stress distribution,

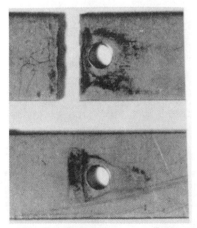

FIGURE 11.9 Failure caused by fretting fatigue [16] (reprinted with permission of R. B. Heywood).

however, may be difficult to model for complex fretting conditions, which implies that these crack growth life predictions may not be reliable.

The total fretting fatigue life consists of the above mentioned crack growth life plus the crack initiation life. The mechanism of fretting that eventually results in localized microcracks is presently conjectural [15-21]. On one hand we have the basic fatigue mechanisms similar to those described in Section 3.3, while on the other we have the ideas of adhesion, abrasion, and corrosion. Regardless, there are certain known phenomea that do occur under fretting conditions.

The contact of two mating surfaces occurs at local high asperities. An oscillatory rubbing action produces tangential cyclic shear stresses, which, along with high Hertzian stresses, can cause local plastic deformation in these asperities. Microwelding and fracture of these asperities can occur and be repeated under the small oscillatory relative motion, causing transfer of metal from one surface to another. Localized high temperatures can also occur, which can accelerate oxidation. The fretting debris that depends on the corrosive environment consists of oxides and metal. The debris is harder than the base metal, which can accelerate abrasion. The volume of the debris is larger than the base particles removed and hence the entrapped accumulated particles can become embedded in the base material and also cause seizure. Environment plays a major role in the fretting process. Both air and humidity contribute to reduced fretting fatigue resistance. Fretting fatigue strengths are higher in vacuum and inert atmosphere. The fretting debris still occurs, in vacuum and inert atmospheres, but the debris particles are the base metal and not oxides. Corrosive environments tend to decrease fretting fatigue resistance and thus fretting fatigue mechanisms are both mechanical and chemical.

Prefretting followed by usual fatigue testing can reduce fatigue strengths as a result of the notch effect caused by pitting. This reduction, however, is substantially less than that under complete fretting fatigue conditions. Thus a conjoint action among fretting, cyclic stressing, and corrosion exists that is similar to corrosion fatigue as shown in Section 11.1.2.

Figure 11.10 shows a shrink or press-fitted shaft in a housing and with a hub, wheel, or gear. Both the unloaded and loaded conditions are shown. It is seen that the major contact area exists at the interface on the compression side of the shaft. It is this compression region where fretting fatigue initiates. If the shafts rotate, then the entire shaft perimeter can be subjected to fretting.

11.2.2 Influence of Variables

The magnitude of the normal pressure between the mating surfaces appreciably affects fretting fatigue strengths. Waterhouse [15] integrated research by others and showed that increases in the normal pressure can produce a substantial decrease in fretting fatigue strength as the pressure increases

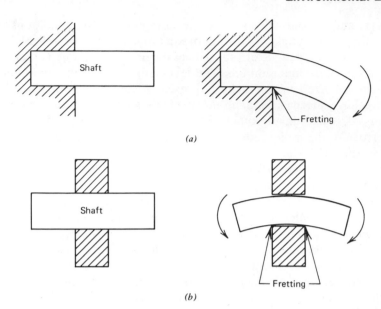

FIGURE 11.10 Deformations of shaft/fixture indicating fretting locations. (*a*) Shrink or press-fit shaft. (*b*) Hub, wheel, or gear on a shaft.

from 0 to about 50 MPa (7 ksi) for aluminum, titanium, nickel, and steel alloys. Pressure increases beyond this value, however, do not have significant additional decreases.

Fenner and Field [19] found that increasing the relative nominal slip amplitude from 0 to 2.5 to 22.5 μm decreased the fatigue strength at 2×10^7 cycles by factors of about 1.7 and 5, respectively in an aluminum alloy.

Lubrication to reduce the coefficient of friction ranges widely in effectiveness depending on the characteristics of the lubricant. Improvement in fretting fatigue strengths due to oils or greases range from zero to a few percent. Dry lubricants such as molybdenum disulfide have produced improvements in fretting fatigue strength up to 20 percent.

Lower temperatures produce greater fretting, which was first determined by noting more damage during the winter than in the summer. Absorbed moisture as a lubricant is important at low temperatures. Quantitative fretting fatigue resistance at low temperature is not clear. At elevated temperatures a higher oxidation rate may exist, but fretting fatigue resistance is also not that well established.

The influence of hardness on fretting fatigue resistance involves the interaction of both mating materials. In general, fretting fatigue resistance decreases with higher hardness, and hence higher strength materials.

Compressive self-stresses at the surface can have a substantial benefit on fretting fatigue resistance. Shot-peening, surface rolling, tensile prestrain, and nitriding can increase fretting fatigue strengths to almost the value of the nonfretted material. Carburizing and surface induction hardening are also beneficial.

Special soft pads or surface coatings between the mating surfaces, such as pure aluminum or magnesium metal shims and Teflon coatings, have increased fretting fatigue resistance. These pads or coatings, however, can be fretted away and must be replaced periodically.

Proper stress transfer design details can also signficantly improve fretting fatigue resistance. For example, the press-fits shown in Fig. 11.10 can have increased fretting fatigue resistance by machining stress relief grooves in the shaft adjacent to the junction as is shown in Fig. 6.13. A reduction in shaft diameter at the junction is also beneficial. The use of tapered plates (both width and thickness) in lap joints has increased fretting fatigue resistance.

In summary, design against fretting fatigue is quite qualitative and trial and error is still required. Accelerated test programs, although a must, can often miss the fretting fatigue failures. Fretting will not often be eliminated, but it can be reduced by judicious alteration of the above described variables. References 15, 17, and 18 can be very helpful in designing against fretting fatigue.

11.2.3 Dos and Don'ts in Design

1. Don't overlook fretting fatigue since it can be quite deleterious and can occur with extremely small ($<10^{-5}$ mm) relative amplitude in most assembled components and structures.
2. Do consider methods of reducing fretting fatigue through the use of compressive self-stresses using, for example, shot-peening, surface rolling, or nitriding.
3. Do consider using softer materials as inserts and coatings between two hard surfaces if possible. These may have to be frequently replaced, however.
4. Don't expect substantial improvements in fretting fatigue strengths from lubricants such as oil and grease. Dry lubricants such as molybdenum sulfide have only been slightly more beneficial.

11.3 LOW TEMPERATURE FATIGUE

Fatigue behavior at low temperatures has received much less attention than that at room temperature and elevated temperatures. This is apparently due to the beneficial effects of low temperature on constant amplitude fatigue behavior found from $S-N$ curves. Mann [14] states

> The fatigue strengths of most metallic materials increase as the temperature is reduced below normal. . . . It appears the rate of increase

in notched fatigue properties of plain carbon steels with decreasing temperature may be less than that of the unnotched properties. . . . Inasmuch as the actual fatigue strengths of most materials at low temperatures do not appear to be less than those at room temperatures, reduced temperatures should not introduce any additional general problems from the fatigue viewpoint.

Munse and Grover [22] state,

However since many structural steels become susceptible to brittle fracture at low temperatures the possibility should not be overlooked of a brittle fracture starting from a small fatigue crack which might otherwise require a much longer time for propagation.

The latter excerpt seems to imply that there should be concern about fatigue failures at low temperatures. This is especially true since the above ideas on low temperature fatigue were primarily based on constant amplitude $S-N$ tests. We consider low temperature fatigue behavior first by reviewing the effect of low temperatures on monotonic material properties and then consider $S-N$, low cycle fatigue, and crack propagation behavior. Most reported low temperature fatigue behavior has been based on these constant amplitude tests and little verification of real-life fatigue predictions have been published for low temperatures.

11.3.1 Monotonic Behavior at Low Temperatures

In general, the unnotched ultimate tensile strength and yield strength increase at lower temperatures for metals with the ratio of the ultimate strength to the yield strength approaching a value of one at lower temperatures. The ductility as measured by the percent elongation or reduction in area at fracture usually decreases with lower temperatures, while the modulus of elasticity usually has a small increase. The total strain energy or toughness at fracture usually decreases at lower temperatures, as measured by the area under the stress–strain curve. Under notched conditions, toughness and ductility can decrease even further. This is true for both low and high strain rates. Impact energy as measured from the Charpy impact (CVN), precracked Charpy (K_{Id}) or dynamic tear (DT) test can show substantial decreases. An upper and lower shelf and transition region usually exist for the low and medium strength steels. Higher strength steels and other metals usually have a more continuous energy curve. Both plane stress fracture toughness, K_c, and plane strain fracture toughness, K_{Ic}, often decrease with lower temperatures. The nil-ductility temperature, NDT, as measured from the drop weight test using a brittle weld bead with a machined notch, has varied from above room temperature to almost absolute zero for steels. Thus it is quite well known that impact energy absorbing capabilities of notched or cracked components can be drastically reduced at lower temperatures depending on microstructure and alloy system. This implies greater notch and crack sensitivity exists at lower temperatures. Final fatigue crack lengths at fracture can then be drastically reduced at lower temperatures. The lower fracture toughness, lower ductility, and higher unnotched tensile strength do not, however, provide sufficient information as to how cracks will initiate and propagate in components under fatigue loadings at low temperatures.

11.3.2 Stress–Life Behavior, S–N

Comprehensive summaries of $S-N$ fatigue behavior at low temperature have been made by Teed [23] in 1950 and by Forrest [4] in 1962. A tabular summary by Forrest for carbon steels, alloy steels, and cast steels is shown in Fig. 11.11. Here the averages of long-life fully reversed fatigue strengths at low temperature divided by the fully reversed fatigue strengths at room temperature are shown for unnotched and notched specimens. No effort was made to correlate strength levels nor stress concentration factors. The goal was to provide a general trend for long-life fatigue strengths at low temperatures compared to room temperature. The number of materials is given at the bottom of each column. From a design standpoint, the most important aspect of Fig. 11.11 is the substantially smaller increases in fatigue strength in the notched specimens.

Figure 11.11 does not give an indication of the complete $S-N$ behavior at low temperature relative to room temperature. Spretnak et al. [24] determined complete $S-N$ behavior of unnotched and notched specimens between 10^3 and 10^7 cycles at low temperatures for many materials. Their results and others can be summarized as follows. At short and long lives, low temperatures are usually beneficial to constant amplitude unnotched $S-N$ fatigue behavior. At shorter lives low temperatures do little good or harm to constant amplitude notched $S-N$ fatigue behavior, while at longer lives notched fatigue strengths are usually just slightly better or similar to room temperature values.

11.3.3 Low Cycle Fatigue, ε–N

Very few low cycle, $\epsilon-N$, fatigue data exist at low temperature. Nachtigall [25] determined low cycle fatigue behavior of 10 different materials using unnotched specimens at room temperature (300 K) and two cryogenic temperatures of 78 (liquid nitrogen) and 4 K (liquid helium). Strain–life curves for five of the materials from Nachtigall are shown in Fig. 11.12. In all 10 cases, at high cyclic fatigue lives, where the elastic strain range component is dominant, the fatigue resistance increased at the cryogenic temperatures. Conversely, at low cyclic lives, where the plastic strain range component is dominant, the fatigue resistance generally decreased with decreasing

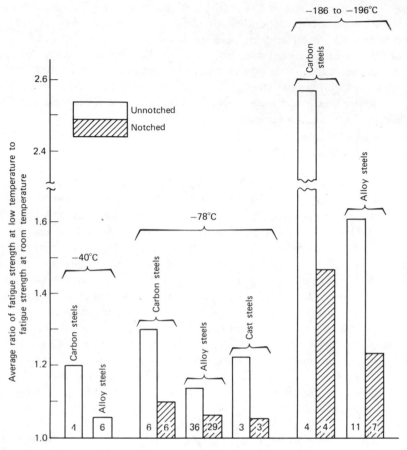

FIGURE 11.11 Comparison of average fatigue changes in fatigue strengths at low temperatures and room temperature for unnotched and notched steels [4] (data used with permission of Pergamon Press Ltd.).

temperature. Only one nickel base alloy, Inconel 718, showed increased fatigue resistance over the entire life range at the cryogenic temperatures. A substantial decrease in fatigue resistance at short lives occurred for the 18Ni maraging steel. This was accompanied by a drastic reduction in ductility, as measured by the percent reduction in area, at 4 K. This great loss in ductility explains the substantial decrease in fatigue resistance at short lives, where the plastic strain range is predominant. All 10 materials had an increase in ultimate tensile strength and a decrease in ductility.

Nachtigall used the Manson method of universal slopes, Eq. 5.13, to predict the strain-cycling fatigue behavior of the 10 materials at cryogenic temperatures with a degree of accuracy similar to that obtained for room temperature results. He concluded that low cycle fatigue behavior of these

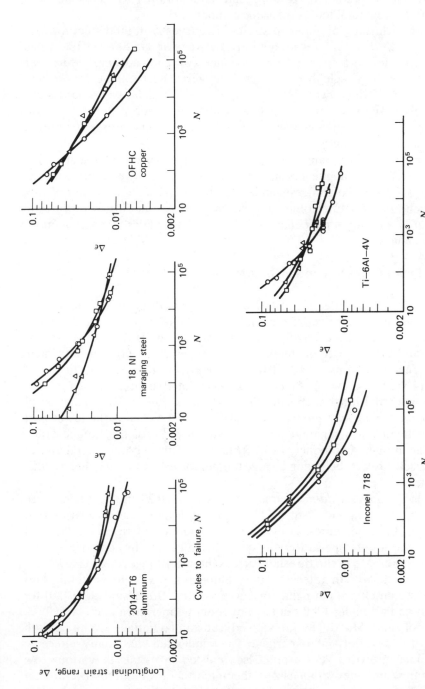

FIGURE 11.12 Effect of cryogenic temperatures on low cycle fatigue behavior [25]. Test temperature, °K: (○) 300 (ambient air), (□) 78 (liquid nitrogen), (△) 4 (liquid helium).

237

materials can be predicted for cryogenic temperatures by using material tensile properties obtained at the same temperatures.

Polák and Klesnil [26] obtained strain–life curves with mild steel for 295, 213, and 148 K. Their data were obtained between about 200 and 10^5 cycles to failure. They found poor fatigue resistance at the lower temperatures for the shorter lives, which they attributed to very short fatigue cracks and brittle fracture. Kikukawa [27] showed that the plastic strain range–life curves between about 5 and 10^3 cycles tend to be lower at lower temperatures. They showed this detrimental effect at low temperatures for both a low and medium strength steel.

A summary of the strain–life low cycle fatigue behavior indicates that unnotched long-life fatigue resistance is increased at lower temperatures, while the short–life fatigue resistance may be decreased as a result of low ductility and low fracture toughness. At short lives, ductility is a principal parameter in strain-control behavior, while at longer lives strength is more the controlling factor.

11.3.4 Fatigue Crack Growth, *da/dN–ΔK*

To make a complete analysis of fatigue crack growth behavior under constant amplitude conditions, the complete sigmoidal curve with the three regions, as shown in Fig. 5.19, must be considered. However, most low temperature and even room temperature fatigue crack growth data have been obtained in regions II and III with crack growth rates greater than 10^{-8} m/cycle (4×10^{-7} in./cycle). Any conclusions related to fatigue crack growth rate are only valid for that specific range of investigation. Extrapolation of curves to the threshold region I is not correct. However, it is the lower part of region II and also region I that account for most of the fatigue crack growth life in many components and structures. Thus a complete picture of fatigue crack growth behavior at low temperature has not yet been sufficiently determined.

Complete sigmoidal da/dN versus ΔK curves at 20 and $-160°C$ were obtained by Yarema and Ostosh [28, 29] using a low carbon steel and an alloy steel. At lower crack growth rates (region I and the lower part of region II) the low temperature was quite beneficial, with approximately a 100 percent increase in the threshold stress intensity range, ΔK_{th}. However, at higher crack growth rates the low temperature was detrimental. This same fatigue crack growth behavior was found by Broek and Rice [30] for regions II and III using 1080 rail steel at room temperature and $-40°C$.

Gerberich and Moody [31] reviewed substantial fatigue crack growth behavior for a variety of materials. They indicated that many alternate fatigue crack growth fracture processes existed in metals at various low temperatures and they emphasized the importance of microstructure on fatigue behavior. They showed that region II fatigue crack growth rates could be substantially altered at different low temperatures. For example,

Fig. 11.13 shows constant amplitude results for an Fe–2.4Si steel with S_y = 200 MPa (29 ksi) at four different temperatures using compact specimens. When the temperature was decreased from 296 (room temperature) to 233 K substantial decreases in da/dN occurred. As the temperature was further decreased to 173 K and then to 123 K the crack growth rates increased for a given stress intensity factor range, with some crack growth rates becoming poorer than room temperature rates. Thus a reversal in region II fatigue crack growth behavior occurred at about 233 K. Electron fractographs revealed that cyclic cleavage became the predominant mode of fatigue crack growth at temperatures below that at which the reversal in behavior occurred. Thus a ductile–brittle fatigue transition temperature exists for these steels. Gerberich and Moody showed that region II slopes could increase sharply for many steels at low temperatures except for some of the Fe–Ni steels. They suggested a parallel exists between a ductile–brittle fatigue

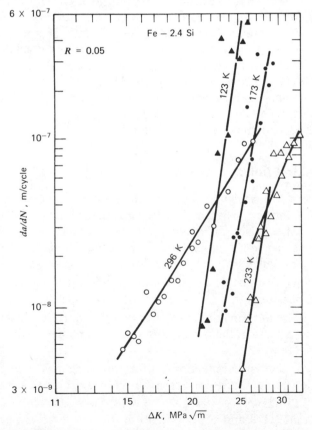

FIGURE 11.13 Fatigue crack growth rate versus ΔK in Fe-2.4Si at room and low temperatures [31] (reprinted by permission of the American Society for Testing and Materials).

transition temperature and the monotonic ductile–brittle transition temperature. Kawasaki *et al.* [32], however, found that the fatigue crack growth transition temperature was substantially below the NDT or CVN transitions. Stonesifer [33] also indicated that CVN ductle–brittle transition temperature mechanisms can be completely different than ductile–brittle transition temperature fatigue crack growth mechanisms. Tobler and Reed [34] showed that Fe–Ni alloys provided similar or better fatigue crack growth resistance as long as the temperature remained in the "upper shelf" range, which was defined as the region where dimpled rupture or fibrous fractures occur during static fracture toughness tests. Cleavage cracking led to drastic acceleration of fatigue crack growth rates at temperatures below the transition region. It has also been determined that when large decreases in fracture toughness occur at low temperatures crack initiation may constitute almost the entire low temperature fatigue life under this condition.

In summary, it appears that the rather positive attitude toward low temperature increased fatigue resistance may be justified with many materials and components. However, sufficient data exist to raise concern about room temperature fatigue design being satisfactory at low temperatures, particularly for short lives and with sharply notched or cracked parts [35]. This is also important since knowledge of low temperature interaction and sequence effects is still rather vague for both crack initiation and crack propagation. It appears reasonable that constant amplitude low temperature fatigue properties should be considered applicable to fatigue life predictions under real-life load histories in a manner similar to that used at room temperature.

11.3.5   Dos and Don'ts in Design

1. Don't adopt the design philosophy that room temperature fatigue life predictions and test programs will be satisfactory for all low temperature conditions. Each material/loading situation should be considered independently since low temperatures can be beneficial or detrimental or have little influence on total fatigue life.

2. Don't believe that just because monotonic tensile strengths increase with lower temperatures that fatigue properties also increase.

3. Do think more in terms of the smaller long-life fatigue strength increases for notched parts rather than the larger unnotched long-life fatigue strength increases.

4. Do consider that large reductions in fracture toughness and ductility can occur at low temperatures and result in very short crack sizes at fracture that might otherwise require a longer time for propagation.

5. Do note that NDT and CVN transition temperatures are generally different than the fatigue transition range, which may be similar or substantially lower. Lower NDT and CVN transition temperatures, however, appear to be accompanied by lower fatigue transition temperatures.

11.4 HIGH TEMPERATURE FATIGUE

Fatigue behavior and life predictions are extremely more complicated at high temperatures than at room temperature. A very complicated interaction between thermally activated time dependent processes is involved. These include environmental, creep/relaxation, and metallurgical aspects in conjoint action with mechanical fatigue mechanisms. Factors such as frequency, wave shape, and creep/relaxation, which are of small consequence at room temperature, have appreciable importance at high temperatures. Time dependent fatigue may thus be a better description at high temperatures. Extrapolation of short-term test results to long-term product requirements is common. The mode of both static and fatigue fracture tends to shift from transcrystalline to intercrystalline as the temperature is raised. This change for fatigue conditions occurs at higher temperatures compared to creep conditions. In general, fatigue resistance for a given metal in an air environment decreases as the temperature increases.

Unequal heating of parts of a component can produce thermal stresses that can lead to fatigue failure. This is called "thermal fatigue" and is discussed by Manson [36]. In this book we assume uniform temperature within a component or specimen.

Oxidation plays a key role in high temperature fatigue and creep. Protective oxide formation is a major factor in the fatigue resistance of a given material. These protective oxide films, however, can be broken down by reversed slip causing much shorter high temperature crack initiation life. Crack propagation rates are also accelerated by high temperature environmental oxidation. Freshly exposed surfaces produced by local plasticity can oxidize rapidly. Grain boundaries are selectively attacked by oxygen. Tests at high temperature in a vacuum or inert atmosphere have shown substantial increases in fatigue/creep resistance compared to high temperature air tests. Thus local oxidation is one of the primary factors in degradation of fatigue/creep resistance at high temperatures. Frequency and wave shape effects are also substantially reduced at high temperatures in a vacuum or inert atmosphere. High temperature fatigue cracks in a vacuum are more frequently transcrystalline, which indicates that oxygen is heavily responsible for the intercrystalline cracks in air.

11.4.1 Stress–Strain Behavior under Cyclic and Hold Times

High temperature load histories often contain hold times at a given stress or strain. Under constant stress conditions creep or crack extension may occur, which results in a change in deformation. Under constant strain conditions relaxation may occur, which results in a reduction of the applied stress. Coffin [37] has summarized the basic stress–strain relationships for various cyclic and hold-time histories as shown in Fig. 11.14. Here it is seen that the hysteresis loops are quite complex and discontinuous. Figure

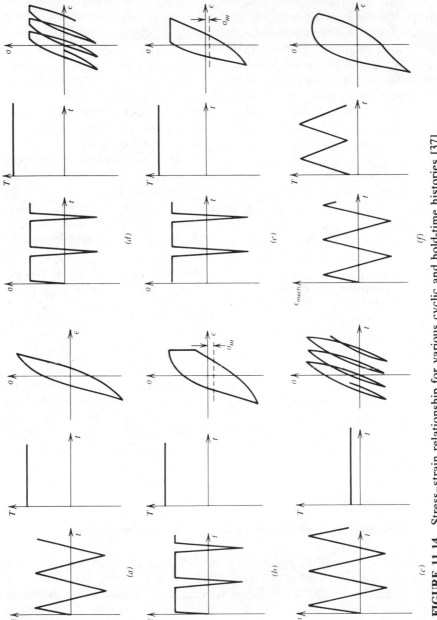

FIGURE 11.14 Stress–strain relationship for various cyclic and hold-time histories [37] (reprinted with permission from *Fracture, 1977: Advances in Research on the Strength and Fracture of Materials*, edited by D. M. R. Taplan, Pergamon Press). (*a*) Continuous strain cycling. (*b*) Strain hold. (*c*) Continuous mean stress cycling. (*d*) Stress hold. (*e*) Stress hold strain limit (CP). (*f*) Mixed mechanical–thermal.

11.14*f* shows both mechanical and thermal mixed cycling. For low cycle fatigue life predictions, Manson [38] partitioned these strains into four inelastic strain ranges that may be used as basic building blocks for any conceivable hysteresis loop. From Fig. 11.14, we can quickly appreciate the difficulty of predicting fatigue life of parts subjected to real-life load histories at high temperature.

11.4.2 Stress–Life/Creep Behavior, S–N

Metals at high temperatures do not usually show a fatigue limit. The fatigue strength continuously decreases with cycles to failure. Thus 10^8 cycles may be a reasonable value for obtaining a long-life fatigue strength. Figure 11.15 obtained by Forrest [4] provides a very comprehensive idea of how the long-life fully reversed fatigue strengths of many materials are influenced by high temperature. The range of temperature is from 20 to almost 1000°C. Data for a given material do not include the entire temperature range because each material has a specific working temperature range over which it is economically and structurally feasible. This is usually a function of melting temperature. The aluminum and magnesium alloys are only applicable at temperatures up to about 200 to 300°C, while the Ni–Cr and Co alloys are predominant at higher temperature between 600 and 900°C. In all cases except for mild steel and cast iron the fatigue strengths decrease with increasing temperature. This anomaly for mild steel and cast iron is due to cyclic strain aging and is accompanied by a decrease in ductility. Fatigue strengths for the temperatures shown in Fig. 11.15 vary by a factor of about 2.5 or less for a given material. If all the data were compared with room temperature fatigue strengths, even larger reductions would occur. Thus degradation of fatigue strengths at reasonable high working temperatures can be quite substantial.

Notches at high temperature are detrimental under predominantly fatigue conditions. However, under predominantly creep conditions notches can either decrease or increase the strength based on net section stresses. Interaction between creep and fatigue can thus provide different notch effects. For example, Vitovec and Lazan [39] determined that net section creep rupture strengths at 900°C (1650°F) in S-816 alloy using notched specimens with $K_t = 3.4$ were better than those for unnotched specimens, as shown in Fig. 11.16*a*, while under fully reversed fatigue conditions the unnotched fatigue strengths were superior, as shown in Fig. 11.16*c*. With stress ratio $A = S_a/S_m = 0.67$ (Fig. 11.16*b*) mixed results occurred. At shorter lives the notch strength was less, while at longer lives it was greater than the unnotched strength. In general, metals are less notch sensitive at high temperatures because of localized plastic and creep flow at notches and the general oxidation of the unnotched or notched surfaces. Self-stresses also have less effect at high temperatures as a result of stress

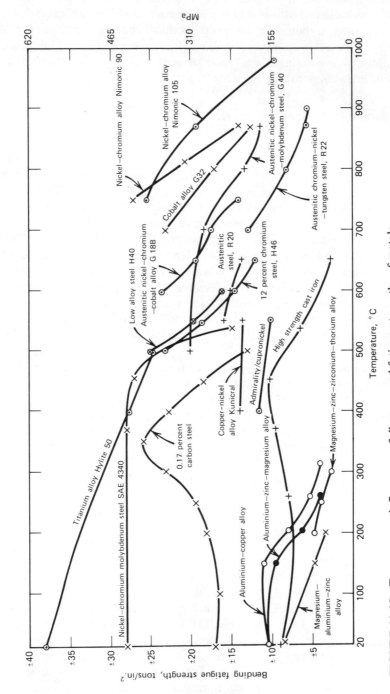

FIGURE 11.15 Temperature influence on fully reversed fatigue strengths of metals [4] (reprinted with permission of Pergamon Press).

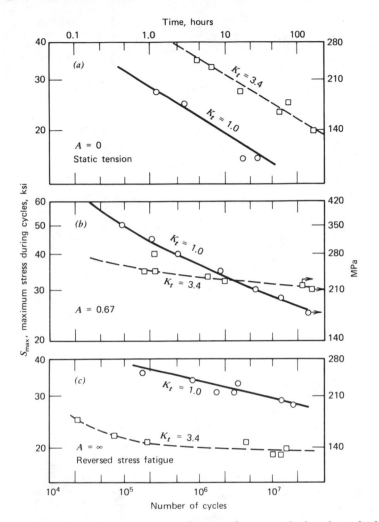

FIGURE 11.16 *S–N* and stress rupture diagrams for unnotched and notched S-816 alloy specimens at 900°C (1950°F) [39].

relaxation from lower yield strengths and plastic and creep flow. However, they can still be beneficial in many situations.

A general effect of tensile mean stress, notches, and creep at high temperature on fatigue as obtained with S-816 alloy by Vitovec and Lazan [39] is shown in Fig. 11.17. These results were obtained under load control at a constant frequency. The curves of Fig. 11.17 are for lives of 2.16 × 10⁷ cycles or 100 hours. Since the tests were performed at constant frequency a direct relationship between cycles and hours exists. The solid lines represent faired-in unnotched data and the dashed lines represent faired-in data for K_t = 3.4. Four different test temperatures in air ranging from room

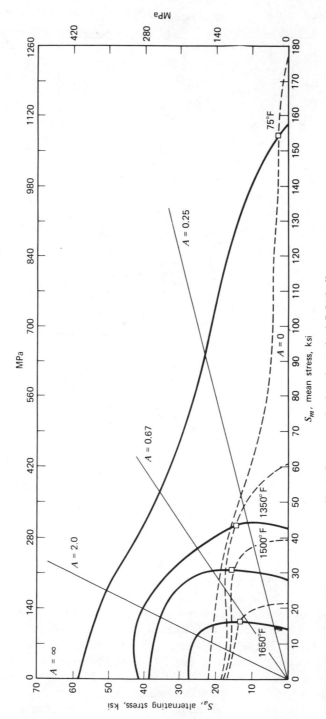

FIGURE 11.17 Tensile mean stress effects for unnotched and notched S-816 alloy specimens for 100 hours life or 2.16 × 10⁷ cycles [39]. (——) Unnotched specimens, $K_t = 1$. (- - -) notched specimens, $K_t = 3.4$.

temperature to 900°C (1650°F) are shown. The vertical axis represents fully reversed fatigue conditions and the horizontal axis represents creep rupture strengths at high temperature and ultimate tensile strength at room temperature. As the temperature is increased both creep rupture strengths and fully reversed fatigue strengths decrease. The curves tend to approximate ellipses or circles as the temperature increases. Thus a first approximation for mean stress effects where both unnotched fatigue and creep are involved is

$$\left(\frac{S_a}{S_f}\right)^2 + \left(\frac{S_m}{S_R}\right)^2 = 1 \qquad (11.1)$$

where S_a = alternating stress
  S_m = mean stress
  S_f = fully reversed fatigue strength
  S_R = creep rupture strength

The modified Goodman curve or the Gerber parabola, Eq. 5.9a and 5.9b, would generally be conservative for high temperature unnotched mean stress effects. They may be unconservative for some notched behavior, however. Substantial increases in S_m can be made before S_a decreases. The open squares represent the intersection of the unnotched and notched behavior for a given temperature. For conditions with high mean stress and thus high creep involvement, notch strengthening is shown. Similar trends occur for other materials and temperatures.

11.4.3 Low Cycle Fatigue, ϵ–N

The gas turbine, steam turbine, and nuclear power fields have created the principal motivation for low cycle high temperature fatigue design information. These are fields where cyclic loads are periodically superimposed on long-term static creep loads. Typical low cycle fatigue behavior obtained in air by Berling and Slot [40] using 304 stainless steel is shown in Fig. 11.18 for three temperatures and two cyclic frequencies. The solid curves in Fig. 11.18 are for 10 cpm and the dashed curves are for 10^{-3} cpm. For a given high temperature, the lower frequency has less fatigue resistance and as the temperature increases the fatigue resistance decreases. Coffin [37, 41] indicates this high temperature low cycle fatigue behavior is typical for air environments and he attributes the frequency and temperature effects to environmental aspects, primarily oxidation.

Figure 11.19, reproduced from Berling and Conway [42], shows the influence of tensile hold time on low cycle fatigue life for 304 stainless steel at 650°C in air. The hold time periods for each cycle are indicated and ranged from 1 minute to 3 hours. As the hold time increased, the life decreased drastically at both test strain ranges as shown. This decreased life due to

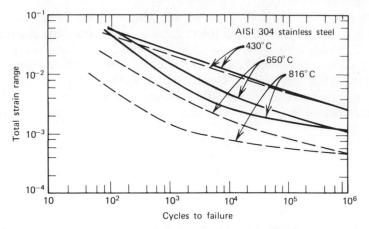

FIGURE 11.18 The effect of temperature and frequency on low cycle fatigue, 304 stainless steel [40] (reprinted by permission of the American Society for Testing and Materials). (——) 10 cpm, (- - -) 10^{-3} cpm.

tensile hold time is generally found in most materials and is more detrimental then equal hold times in tension/compression or compression only [37].

Application of low cycle fatigue behavior to design of notched components at high temperature has been used with the notch strain approach as is described in earlier chapters for life to short cracks. Shortcomings exist, however, as a result of creep/fatigue/environment interaction. Manson [43] established one of the early modifications for high temperature use by reducing the basic room temperature equation (Eq. 5.10 or 5.13) by a 10 percent rule. The rationale involved was to account for the accelerated crack initiation and propagation stages brought about by intercrystalline

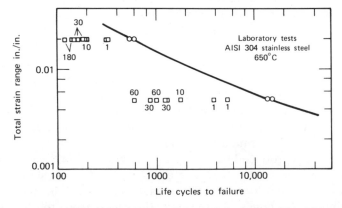

FIGURE 11.19 The effect of hold time on low cycle fatigue life, 304 stainless steel at 650°C]42]. (○) No hold time, (□) tensile hold time in minutes as indicated.

deterioration. The right side of the universal slopes equation (Eq. 5.13) developed for room temperature was divided by a factor of 10 to account for this acceleration. Coffin [44] has modified Eq. 5.10 by introducing empirical frequency factors to allow for different frequency effects. The strain range partitioning method indicated earlier has been used in several high temperature low cycle fatigue situations including combustion liners on jet engines [45]. Damage effects under combined creep/fatigue conditions have been calculated by linearly summing the fractions of damage due to creep with that of fatigue such that

$$\sum \text{fatigue damage} + \sum \text{creep damage} = 1 \qquad (11.2)$$

11.4.4 Fatigue Crack Growth, da/dN–ΔK

As at other temperatures, the stress intensity factor range ΔK and the ratio R can describe high temperature fatigue crack growth behavior. As temperatures are increased the fatigue crack growth rates increase for a given ΔK as shown by James [46] in Fig. 11.20 for Hastelloy X-280. Here it is seen that da/dN, for a given ΔK is more than an order of magnitude higher at 649°C than at room temperature. All tests were run at a given frequency and in an air environment. Threshold levels ΔK_{th} also decrease with higher temperatures. James and Knecht [47] have shown that propagation of fatigue cracks at high temperature in a vacuum or inert atmosphere can be significantly less than in air and very similar to that at room temperature. They also attribute the greater fatigue crack growth rate at high temperature to environmental (oxidation) effects.

Typical frequency effects at high temperature on fatigue crack growth rates obtained by James [48] for 304 austenitic stainless steel at 538°C are shown in Fig. 11.21. Frequencies varying from 0.083 to 4000 cpm are shown and more than an order of magnitude difference exists for these data between the two extreme frequencies. Again this is attributed primarily to environmental effects. The Paris equation (Eq. 5.17) has been commonly used for region II crack growth behavior for high temperature such that

$$\frac{da}{dN} = [C(f)](\Delta K)^n \qquad (11.3)$$

where $C(f)$ is a function of frequency and temperature.

Mean stresses alter fatigue growth rates as shown, for example, in Fig. 11.22 for Inconel 600 obtained by James [49] at 427°C. At higher R ratios the fatigue crack growth rates are higher for a given ΔK value. The largest differences appear at threshold levels.

Fatigue crack growth life predictions at high temperatures can be made by integrating the crack growth rate data as in Chapter 8 and 10 for room temperature. Much less is known, however, about interaction effects, plastic zone sizes, and retardation.

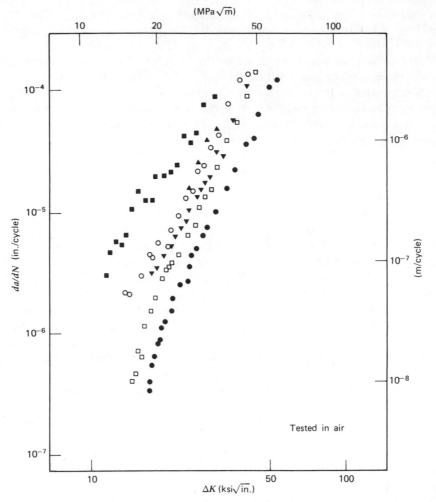

FIGURE 11.20   The effect of temperature on fatigue crack growth behavior, Hastelloy X-280, $R = 0.05$ [46]. Temperature, °C: (●) 24, (□) 316, (▼▲) 427, (○) 538, (■) 649.

11.4.5   Dos and Don'ts in Design

1. Do consider ways of protecting the surface from air at high temperatures, since oxidation is one of the principal causes of fatigue resistance degradation at high temperatures.
2. Don't assume that notch sensitivity is substantially reduced at high temperatures. Reduce stress concentrations if possible.
3. Do consider self-stresses for increased high temperature fatigue resist-

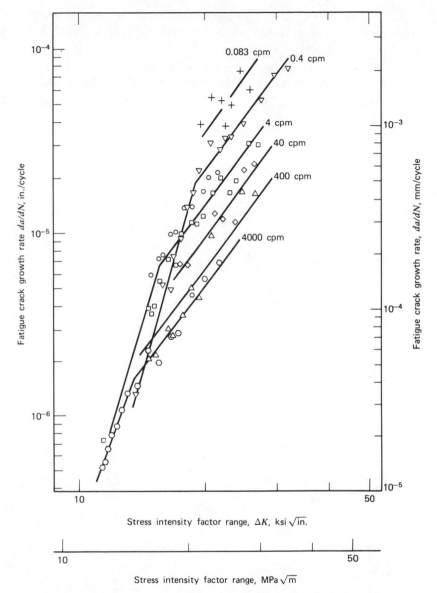

FIGURE 11.21 The effect of frequency on fatigue crack growth behavior, 304 stainless steel at 538°C (1000°F), $R = 0.05$ [48] (reprinted by permission of the American Society for Testing and Materials).

ance, but remember they can have less influence at high temperature than at room temperature because of possible high temperature stress relaxation from plastic strains.

4. Do consider both crack initiation life and crack growth life in considering the entire high temperature fatigue/creep interaction life.

5. Don't neglect thermal stresses if temperatures are not uniform in a part or component.

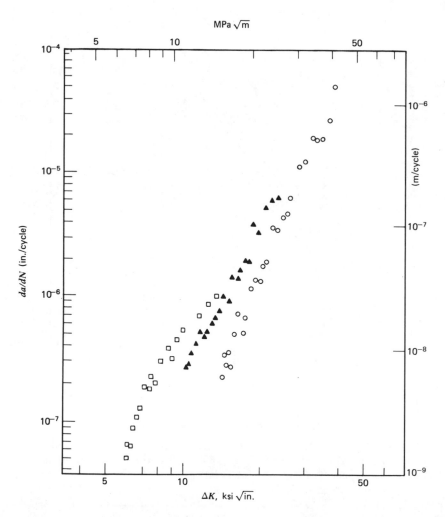

FIGURE 11.22   The effect of R ratio on fatigue crack growth behavior, annealed Inconel 600 in air at 427°C [49]. (○) $R = 0.05$, (▲) $R = 0.333$, (□) $R = 0.6$.

11.5 NEUTRON IRRADIATION

The nuclear reactor field has the greatest interest in neutron irradiation. Here long term steady-state creep loadings at elevated temperatures plus repeated loadings are superimposed on irradiation. Many variables are involved that affect the mechanical properties of reactor materials, namely, neutron fluence, neutron flux, time, irradiation temperature, operating temperature, prior thermomechanical treatment, and basic microstructure. The most common parameter for measuring irradiation effects is the neutron fluence, which is the neutron flux integrated over the exposure time and given in units of neutrons/cm² for energies greater than 0.1 MeV.

The general effect of increasing neutron irradiation fluence on monotonic properties is to increase the ultimate tensile strength and yield strength while decreasing elongation and reduction in area at fracture. The fracture toughness K_{Ic} and the Charpy V notch (CVN) energy usually decrease. The nil-ductility temperature (NDT) for ferritic steels is increased along with an accompanying decrease in upper shelf CVN energy. At these higher strain rates one finds even greater undesirable embrittlement. The effect of neutron irradiation on fatigue properties is not sufficiently known. This is due to past greater emphasis on creep resistance and the very complex nature and expense of fatigue tests of irradiated parts and specimens. Beesten and Brinkman [50] and Brinkman et al. [51] compared low cycle high temperature (400 to 700°C) strain–life behavior for unirradiated and irradiated 304 and 316 austenitic stainless steels. Fatigue lives between 400 and 20,000 cycles were determined. They found that irradiation reduced fatigue life by factors between 1.5 and 2.5.

James [52, 53] has made comprehensive reviews of fatigue crack growth behavior for metals under various neutron fluences, irradiation temperatures, test temperatures, and frequency. The materials included ferritic steels (ASTM alloys A302B, A533B, A508, and A543), austenitic stainless steels (304, 308 and 316), and nickel base alloys (Inconel 625, Inconel 718 and Nimonic alloys). These materials have received the most irradiation fatigue attention. Most fatigue crack growth tests had growth rates higher than 10^{-7} m/cycle (4×10^{-6} in./cycle), with only a few tests with data slightly less than 2.5×10^{-8} m/cycle (10^{-6} in./cycle). Thus all the data reviewed by James were in regions II and III of the sigmoidal da/dN versus ΔK curve. No threshold values were reported. Conclusions, therefore, cannot be made for the entire fatigue crack growth region.

Figure 11.23 reproduced from James [53], shows crack growth rate results for irradiated and unirradiated A533B steel tested at 288°C for various frequencies (10 to 600 cpm) from three different irradiation experiments. Much of the scatter can be attributed to frequency effects in air at high temperature as shown in Section 11.4. For a given frequency, the irradiated

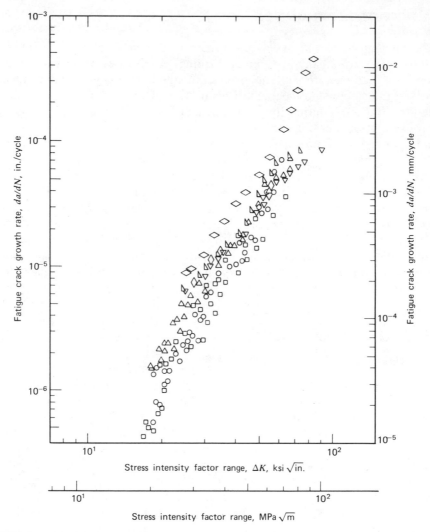

FIGURE 11.23   Fatigue crack growth behavior of irradiated and unirradiated ASTM A533 Grade B Class 1 steel tested in air at 288°C (550°F) [53]. (∇)-unirradiated, others -irradiated.

crack growth resistance was both slightly better and slightly worse than for the unirradiated material. This type of small difference was found for most of the ferritic steels. Both irradiated and unirradiated crack growth resistance decreased at higher temperatures. James concluded for the ferritic steels in regions II and III that constant amplitude fatigue crack growth resistance after irradiation was neither significantly less nor significantly greater than in air environment alone. For irradiated austenitic steels James

[53] found mixed results, with some studies indicating no difference, some studies showing higher growth rates in irradiated material, and others showing just the opposite effect. Thus no real final generalization concerning total fatigue behavior under neutron irradiation can be made at this time. It appears, however, that any constant amplitude fatigue degradation is not as severe as the embrittlement caused by the irradiation. Data on fatigue interaction effects with irradiation under spectrum loading are not available.

REFERENCES FOR CHAPTER 11

1. *Damage Tolerant Handbook,* Metals and Ceramics Information Center, Battelle, Columbus, OH, 1975.

2. M. H. Peterson, B. F. Brown, R. L. Newbegin, and R. E. Grover, "Stress Corrosion Cracking of High Strength Steels and Titanium Alloys in Chloride Solutions at Ambient Temperature," *Corrosion,* Vol. 23, 1967, p. 142.

3. D. J. McAdam, "Corrosion Fatigue of Metals," *Trans., Am. Soc. Steel Treating,* Vol. 11, 1927, p. 355.

4. P. G. Forrest, *Fatigue of Metals,* Pergamon Press, London, 1962.

5. *Corrosion Fatigue: Chemistry, Mechanics and Microstructure,* International Corrosion Conference Series, Vol. NACE-2, National Association of Corrosion Engineers, Houston, 1962.

6. *Corrosion-Fatigue Technology,* ASTM STP 642, 1978.

7. C. Q. Bowles, "The Role of Environment, Frequency and Wave Shape During Fatigue Crack Growth in Aluminum Alloys," Delft University of Technology, Department of Aerospace Engineering, Report LR-270, May 1978.

8. T. W. Crooker, F. D. Bogar, and W. R. Cares, "Effects of Flowing Natural Seawater and Electrochemical Potential on Fatigue-Crack Growth in Several High-Strength Marine Alloys," *Corrosion-Fatigue Technology,* ASTM STP 642, 1978, p. 189.

9. O. Vosikovsky, "Fatigue-Crack Growth in an X-65 Line-Pipe Steel at Low Cyclic Frequencies in Aqueous Environments," *Trans. ASME, J. Eng. Mater. Technol.,* October 1975, p. 298.

10. E. J. Imhof and J. M. Barsom, "Fatigue and Corrosion-Fatigue Crack Growth of 4340 Steel at Various Yield Strengths," *Progress in Flaw Growth and Fracture Toughness Testing,* ASTM STP 536, 1973, p. 182.

11. H. P. Chu and J. G. Macco, "Corrosion Fatigue of 5456-H 117 Aluminum Alloy in Saltwater," *Corrosion-Fatigue Technology,* ASTM STP 642, 1978, p. 223.

12. R. J. Bucci, "Environment Enhanced Fatigue and Stress Corrosion Cracking of a Titanium Alloy Plus a Simple Model for the Assessment of Environmental Influence on Fatigue Behavior," Ph.D. Dissertation, Lehigh University, Bethlehem, PA 1970.

13. J. T. Ryder, W. E. Krupp, D. E. Pettit, and D. W. Hoeppner, "Corrosion-

Fatigue Properties of Recrystallization Annealed Ti–6Al–4V," *Corrosion-Fatigue Technology,* ASTM STP 642, 1978, p. 202.

14. J. Y. Mann, *Fatigue of Materials,* Melbourne University Press, Australia, 1967.

15. R. B. Waterhouse, *Fretting Corrosion,* Pergamon Press, London, 1972.

16. R. B. Heywood, *Designing Against Fatigue of Metals,* Reinhold Publishing Corp., New York, 1962.

17. *Fretting in Aircraft Systems,* AGARD Conference Proceedings No. 161, Agard, Munich, Germany, October 1974.

18. *Control of Fretting Fatigue,* National Materials Advisory Board, Publication NMAB-333, Washington, D.C., 1977. Available as AD-A044 652 National Technical Information Service.

19. A. J. Fenner and J. E. Field, "Fatigue under Fretting Conditions," *Rev. Metall.,* Vol. 55, 1958, p. 475.

20. D. W. Hoeppner and G. L. Goss, "Metallographic Analysis of Fretting Fatigue in Ti–6Al–4V MA and 7075-T6 Aluminum," *Wear,* Vol. 27, 1974, p. 175.

21. J. E. Bowers, N. J. Finch, and A. R. Goreham, "The Prevention of Fretting Fatigue in Aluminum Alloys," *Proc. Inst. Mech. Eng. Lond.,* Vol. 182, Part I, 1967–1968, p. 703.

22. W. W. Munse and L. Grover, *Fatigue of Welded Structures,* Welding Research Council, New York, 1964.

23. P. L. Teed, *The Properties of Metallic Materials at Low Temperatures,* Chapman and Hall, 1950.

24. J. W. Spretnak, M. G. Fontana, and H. E. Brooks, "Notched and Unnotched Tensile and Fatigue Properties of Ten Alloys at 25 and −196°C," *Trans. ASM,* Vol. 43, 1951, p. 547.

25. A. J. Nachtigall, "Strain-Cycling Fatigue Behavior of Ten Structural Metals Tested in Liquid Helium (4 K), in Liquid Nitrogen (78 K) and in Ambient Air (300 K)," NASA TN D-7532, Feb. 1974.

26. J. Polák and M. Klesnil, "The Dynamics of Cyclic Plastic Deformation and Fatigue Life of Low Carbon Steel at Low Temperatures," *Mater. Sci. Eng.,* Vol. 26, No. 2, December 1976, p. 157.

27. M. Kikukawa, M. Jono, T. Kamato, and T. Nakano, "Low Cycle Fatigue Properties of Steels at Low Temperatures," *Proc. 13th Jap. Congr. Mater. Res.,* 1970, p. 69.

28. S. Ya Yarema, "Growth of Fatigue Cracks in Low Carbon Steel under Room and Low Temperatures," *Probl. Prochn.,* No. 3, 1977, p. 21 (in Russian).

29. S. Ya Yarema and O. P. Ostosh, "A Study of Fatigue Crack Growth in Low Temperatures," *Fiz.-Khim. Mekh. Mater.,* No. 2, 1977, p. 48, (in Russian).

30. D. Broek and R. C. Rice, "Prediction of Fatigue-Crack Growth in Railroad Rails," Battelle Columbus Laboratories, 1978.

31. W. W. Gerberich and N. R. Moody, "A Review of Fatigue Fracture Topology Effects on Threshold and Kinetic Mechanism," *Symposium on Fatigue Mechanisms,* J. T. Fong, Ed., ASTM, STP 675, 1979, p. 451.

32. T. Kawasaki, T. Yokobori, Y. Sawaki, S., Nakanishi, and H. Izumi, "Fatigue Fracture Toughness and Fatigue Crack Propagation in 5.5% Ni Steel at Low

Temperature," *Fracture, 1977, ICF-4,* Vol. 3, Waterloo, Canada, June 1977, p. 857.

33. F. R. Stonesifer, "Effect of Grain Size and Temperature on Fatigue Crack Propagation in A533 B Steel," *Eng. Fract. Mech.* Vol. 10, 1978, p. 305.

34. R. L. Tobler and R. P. Reed, "Fatigue Crack Growth Resistance of Structural Alloys at Cryogenic Temperature," paper presented at the Cryogenic Engineering Conference/International Cryogenic Materials Conference, University of Boulder, Colorado, August 1977.

35. R. I. Stephens, J. H. Chung, and G. Glinka, "Low Temperature Fatigue Behavior of Steels—A Review," paper presented at the SAE Earthmoving Conference, Peoria, Illinois, SAE paper no. 790517, April 1979.

36. S. S. Manson, *Thermal Stress and Low Cycle Fatigue,* McGraw-Hill Book Co., New York, 1966.

37. L. F. Coffin, "Fatigue at High Temperature," *Fracture, 1977, ICF-4* Vol. 1, Waterloo, Canada, 1977, p. 263.

38. S. S. Manson, "The Challenge to Unify Treatment of High Temperature Fatigue—A Partisan Proposal Based on Strain Range Partitioning," *Fatigue at Elevated Temperatures,* ASTM STP 520, 1973, p. 744.

39. F. H. Vitovec and B. J. Lazan, "Fatigue, Creep, and Rupture Properties of Heat Resistant Materials," WADC Technical Report No. 56–181, August 1956.

40. J. T. Berling and T. Slot, "Effect of Temperature and Strain Rate on Low-Cycle Fatigue Resistance of AISI 304, 316, and 348 Stainless Steels," *Fatigue at High Temperatures,* ASTM STP 459, 1969, p. 3.

41. L. F. Coffin, "Fatigue at High Temperatures," *Fatigue at Elevated Temperatures,* ASTM STP 520, 1973, p. 5.

42. J. T. Berling and J. B. Conway, "Effect of Hold-Time on the Low-Cycle Fatigue Resistance of 304 Stainless Steel at 1200°F," *1st Int. Conf. Pressure Vessel Technol., Part 2, Delft, Holland,* 1969, p. 1233.

43. S. S. Manson, "Interfaces between Fatigue, Creep, and Fracture," *Int. J. Fract. Mechan.,* Vol. 2, No. 1, 1966, p. 327.

44. L. F. Coffin, "The Effect of Frequency on the Cyclic Strain and Low Cycle Fatigue Behavior of Cast Udimet 500 at Elevated Temperature," *Metall. Trans.,* Vol. 2, 1971, p. 3105.

45. G. R. Halford, and A. J. Nachtigal, "The Strainrange Partitioning Behavior of a Gas Turbine Disk Alloy, AF2-1DA," paper presented at the AIAA/SAE/ASME 15th Joint Propulsion Conference, Las Vegas, June, 1979.

46. L. A. James, "The Effect of Temperature Upon the Fatigue–Crack Propagation of Hastelloy X-280," Report HEDL-TME 76-40, 1976.

47. L. A. James and R. L. Knecht, "Fatigue-Crack Propagation Behavior of Type 304 Stainless Steel in a Liquid Sodium Environment," *Metall. Trans. A,* Vol. 6A, Jan. 1975, p. 109.

48. L. A. James, "The Effect of Frequency upon the Fatigue-Crack Growth of Type 304 Stainless Steel at 1000 F," *Stress Analysis and Growth of Cracks,* ASTM STP 513, 1972, p. 218.

49. L. A. James, "Fatigue-Crack Propagation Behavior of Inconel 600," Report HEDL-TME 76-43, 1976.

50. J. M. Beeston and C. R. Brinkman, "Axial Fatigue of Irradiated Stainless Steels Tested at Elevated Temperatures," *Irradiation Effects on Structural Alloys for Nuclear Reactor Applications,* ASTM STP 484, 1970, p. 419.

51. C. R. Brinkman et al., "Influence of Irradiation on the Creep/Fatigue Behavior of Several Austenitic Stainless Steels and Incoloy 800 to 700°C," *Effect of Radiation on Substructure and Mechanical Properties of Metals and Alloys,* ASTM STP 529, 1973, p. 473.

52. L. A. James, "Fatigue Crack Propagation in Neutron-Irradiated Ferritic Pressure-Vessel Steels," *Nuclear Safety,* Vol. 18, No. 6, November–December 1977, p. 791.

53. L. A. James, "Effects of Irradiation and Thermal Aging upon Fatigue-Crack Growth Behavior of Reactor Pressure Boundary Materials," HEDL-SA-1663, also in *Proc. Int. At Energy Agency,* Technical Committee Meeting, Innsbruck, Austria, Nov. 1978.

PROBLEMS FOR CHAPTER 11

1. A mild steel circular stepped shaft is subjected to pure bending. The shaft diameters are 25 and 15 mm and the notch root radius is 4 mm. The shaft is to operate in flowing seawater for six months. What fully reversed bending moment would you recommend for a life of 10^7 cycles? What moment would you recommend if the shaft were isolated from the water? How would you verify your decisions?

2. What materials or operations would you recommend for problem 1 to increase the fully reversed bending moment by 80 percent?

3. Solve problem 7 in Chapter 8 if the component is to be operated outdoors in a rain forest. Discuss your assumptions and indicate what type of errors you may have introduced.

4. Propose numerical room temperature fully reversed fatigue strengths ($>10^7$ cycles) for mild steel shafts shown in Figs. 11.10*a* and 11.10*b*.

5. Solve problems 7*a* and 7*b* of Chapter 8 assuming the component is to be operated in liquid nitrogen (78 K) and the material is 18 Ni maraging steel as shown in Fig. 11.12.

6. Solve problem 1, Chapter 10, if the component is to operate at −40°C. Discuss the significance of your assumptions and the accuracy of your results.

7. Discuss the ideas of problem 11, Chapter 10, for temperatures of −40°C and liquid nitrogen temperature (78 K).

8. Determine the effect of temperature on K_f in Fig. 11.17. Consider all four temperatures and $A = \infty$, 2.0, and 0.25.

9. Solve problems 7*a* and 7*b* of Chapter 8 assuming the component is annealed 304 stainless steel operating at 7 Hz and 500°C. Comment on the accuracy of your prediction.

CHAPTER **12**

JOINTS

Parts and structures are invariably joined together in some fashion. Common joint types include welded, bolted or riveted, and pinned connections. These joints present difficulties because of stress concentrations, self-stresses, fretting, and misalignment, all of which may vary between nominally equal parts. Joints are frequently the prime location for fatigue failures. We consider these different joints from a fatigue design viewpoint. Since the complexities are even more severe than those indicated in many of the preceding chapters, this again implies that good fatigue design must include testing, inspection, service monitoring, and carefully analyzed experience.

12.1 WELDS

Welding itself is an art that can result in extremely different fatigue resistance. The quality of workmanship and design determines the fatigue resistance of weldments. A carefully designed and processed weldment can develop the same fatigue strength as a part forged and machined from one piece, and at far less cost. An aircraft part that incorporates a nose wheel spindle and a landing gear piston in one piece of high strength steel may serve as an example of a carefully designed weldment. The parts are rough machined from a material like 4340, welded together, finish machined, heated in a controlled atmosphere, quenched in oil, tempered to achieve the desired hardness, and shot-peened. By contrast, an attachment using an untreated fillet weld can reduce fatigue resistance substantially.

We consider welds by indicating the different macro and micro defects that can exist, typical fatigue behavior, and methods for improving fatigue resistance in weldments, and then we discuss the four fatigue design procedures outlined in Chapter 8, for application to weldments.

12.1.1 Weldment Nomenclature and Defects

Several typical weldments are shown schematically in Fig. 12.1 with accompanying design fatigue strengths for structural steels at 2×10^6 cycles with

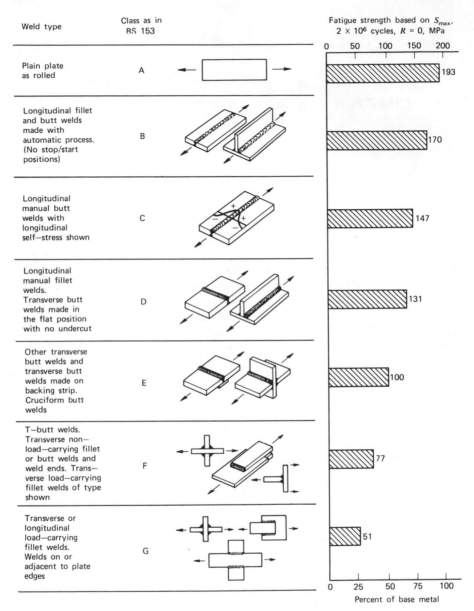

Weld type	Class as in BS 153		Fatigue strength based on S_{max}, 2×10^6 cycles, $R = 0$, MPa
Plain plate as rolled	A		193
Longitudinal fillet and butt welds made with automatic process. (No stop/start positions)	B		170
Longitudinal manual butt welds with longitudinal self—stress shown	C		147
Longitudinal manual fillet welds. Transverse butt welds made in the flat position with no undercut	D		131
Other transverse butt welds and transverse butt welds made on backing strip. Cruciform butt welds	E		100
T—butt welds. Transverse non—load—carrying fillet or butt welds and weld ends. Trans—verse load—carrying fillet welds of type shown	F		77
Transverse or longitudinal load—carrying fillet welds. Welds on or adjacent to plate edges	G		51

Percent of base metal

FIGURE 12.1 Weld type and fatigue strength for structural steel based on BS153 [2], taken from reference [1].

$R = 0$. These values were formulated by Richards [1] based on British welding standards [2]. These are rather simplified joints, but they do simulate many real parts and structures. In general we have butt, fillet, and spot welds with many different weldment shapes and configurations. Both transverse butt and longitudinal butt welds are common. Fillet welds, however, are even more common and may be load carrying as in the left of class G or non-load carrying as in the left of class F. The design fatigue strengths given in Fig. 12.1 range from about 25 to 90 percent of the unnotched base metal strength and provide a reasonable guide for actual usage.

Because of nonuniform temperature gradients and local elastic/plastic deformations during the welding and cooling process, biaxial or triaxial self-stresses are formed in all welds. The self-stress profiles and their magnitudes are difficult to quantify. Figure 12.1 class C shows a typical longitudinal self-stress distribution at a transverse section The self-stresses are tensile in the weld region, which must be balanced by compressive self-stresses away from the weld. The tensile stress may reach values equal to the yield strength and can appreciably contribute to the poor fatigue behavior of some weldments. We consider self-stresses in weldments in more detail in Section 12.1.3.

A polished and etched photograph of a longitudinal section of a cruciform fillet weldment obtained by Albrecht [3] is shown in Fig. 12.2. Three regions

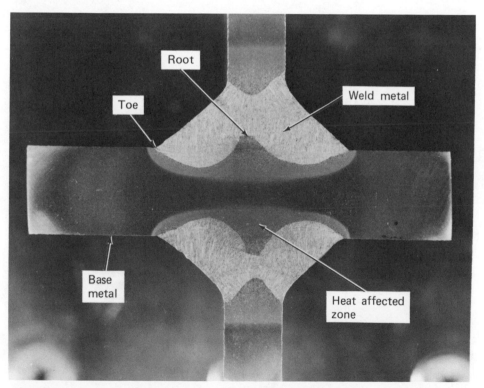

FIGURE 12.2 Polished and etched longitudinal section through a non-load carrying cruciform weldment [3] (reprinted by permission of the American Society for Testing and Materials).

of a weldment are quite clear:

1. Parent or base metal (BM).
2. Weld metal (WM).
3. Heat affected zone (HAZ).

These regions have different microstructure, self-stresses, defects, and mechanical properties. The weld metal is similar to a cast metal with substantial anisotropy. The heat affected zone is the base metal, which is subjected to sufficiently high temperatures during welding to cause recrystallization. It can have different properties than the original base metal.

Figure 12.3 shows transverse sections of full and partial penetration butt and fillet weldments. Common locations for cracks to initiate are shown for each weldment. Stress concentrations which occur at the toe of butt and fillet welds, are common locations for fatigue cracks to initiate. The toe is also at the surface, where bending stresses are the largest. K_t at the toe depends on the geometry of the weld as defined in Fig. 12.3b. It is usually larger in fillet welds than in butt welds. Values of K_t have ranged from

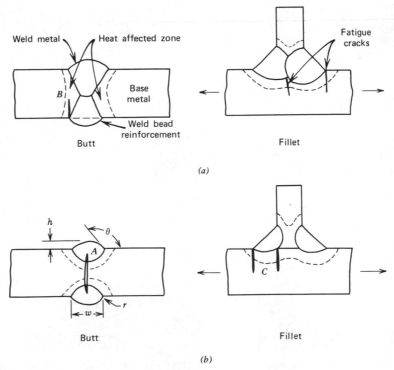

(a)

(b)

FIGURE 12.3 Weldment nomenclature and fatigue crack initiation sites. (a) Full penetration welds. (b) Partial penetration welds.

essentially 1.0, for butt welds with the reinforcement removed, up to values of 3 to 5, for sharp discontinuities in fillet welds. The word "reinforcement" is a misnomer since it implies a positive effect. Actually the "reinforcement" is quite detrimental in fatigue and should perhaps be called "excess" weld metal. Fatigue cracks also initiate at the weld root as is shown in Fig. 12.3 for fillet welds and partial penetration butt welds. In addition, there are many defects in the weldment that can cause cracks to initiate from these. Defects in welds can include slag or oxide inclusions, shrinkage cracks, lack of fusion, porosity, poor penetration, overlaps, surface ripples, undercuts, stray arc strikes, and reinforcement geometry. All these affect fatigue crack initiation, propagation, and total life. Figure 12.3 indicates that fatigue cracks may grow in the weld metal (A) or heat affected zone (B), or through the heat affected zone and the base metal (C). Thus we should have information about fatigue behavior in all three zones.

12.1.2 Constant Amplitude Fatigue Behavior of Weldments

Most steels used in weldments have yield strengths below 700 MPa (100 ksi). Even with strengths beyond this value, much information indicates that for a given transverse butt or fillet joint, as-welded constant amplitude fatigue strengths at 10^6 cycles or more are rather independent of material. This was shown by Reemsnyder [4] for transverse butt weldments using a number of steels in different conditions with tensile strengths varying from 400 to 1030 MPa (58 to 148 ksi). The geometric notch severity of the weldment, loss of heat treatment, decarburization, and self-stresses, which may or may not relax out, are the causes of this behavior. As we see later, however, methods do exist to appreciably alter weldment fatigue resistance.

The two largest factors affecting weldment fatigue resistance are geometrical stress concentrations and self-stresses. The influence of reinforcement angle θ for transverse butt welds is shown in Fig. 12.4 [1]. Here we see that a factor of almost 2:1 exists for the fatigue strength at 2×10^6 cycles as the reinforcement angle varies from 100 to 150 degrees. Reemsnyder [4] showed that decreasing the height h of the transverse butt weld reinforcement continually improved the fatigue strengths of 785 MPa (114 ksi) ultimate strength steel as shown in Fig. 12.5. Fatigue notch factors, K_f, at 2×10^6 cycles were 4.1, 2.2, 1.5, and essentially 1.0 as the height was decreased from 3.8 mm to complete reinforcement removal. Thus S_f for the transverse butt weldment with reinforcing removed is very similar to that for the base material. Sanders and Lawrence [5] also showed significant improvement in fatigue strength in aluminum alloy transverse butt welds when the reinforcement was removed. Thus it should be clear that reducing stress concentrations in weldments is a major factor in improving fatigue resistance.

Fatigue cracks can initiate or grow in the base metal, weld metal, or heat affected zone; however, only limited cyclic stress–strain data and low cycle

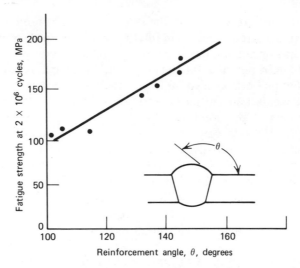

FIGURE 12.4   Influence of reinforcement shape on fatigue strength of transverse butt welds [1].

fatigue data exist for the three weldment regions. Higashida et al. [6] have determined these properties for A36 steel, A514 steel, and 5083 aluminum alloy. Cyclic softening was predominant in the A36 steel WM and HAZ metal. Even greater cyclic softening occurred in the A514, WM, HAZ, and BM metal. For the aluminum alloy, cyclic hardening occurred in the WM

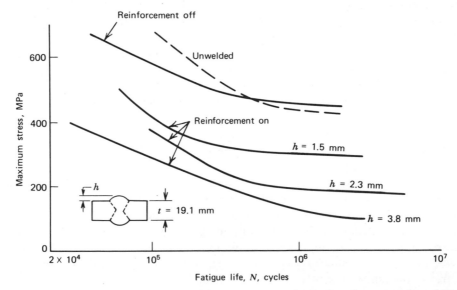

FIGURE 12.5   *S–N* curves for transverse butt welds, Q & T carbon steel, $S_u = 785$ MPa (114 ksi), $R = 0$ [4] (reprinted by permission of the American Society for Testing and Materials).

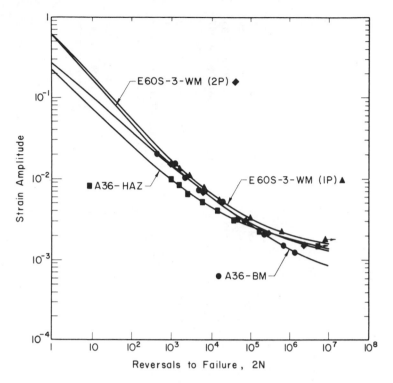

FIGURE 12.6 Strain-controlled fatigue behavior of A36 steel weldment materials [6] (reprinted with permission of the American Welding Society).

and BM metal. Strain–life (ϵ–N) curves for the A36 steel weldment materials are shown in Fig. 12.6 and monotonic and cyclic strain, and strain–life properties for the different materials are given in Table 12.1. The general trend found for the ϵ–N curves for the three weldment regions is that the softer materials have better fatigue resistance at the low lives while the harder materials are better at the longer lives. This agrees with previous ϵ–N data as described in Fig. 5.18.

Maddox [7] obtained fatigue crack growth data for weld metals with S_y ranging from 386 MPa (56 ksi) to 636 MPa (92 ksi), simulated HAZ and C–Mn base metal as shown in Fig. 12.7. Eleven different conditions are superimposed in this figure. The scatter band for all stress intensity factor ranges varied between 2:1 and 3:1 with crack growth rates between 10^{-8} and 3×10^{-5} m/cycle (4×10^{-7} and 1.2×10^{-4} in/cycle). This small scatter band implies that region II fatigue crack growth behavior is similar in sound weldments. James [8] has also indicated this same behavior occurs in pressure vessel steel weldments.

Mean stress influence on fatigue resistance of weldments is similar to other notched components. A constant-life diagram with S_a versus S_m for

TABLE 12.1 Monotonic and Cyclic Strain Properties of Weld Materials: SI units [6]

	S_u, (MPa)	S_y/S_y', (MPa/MPa)	K/K', (MPa/MPa)	n/n'	ϵ_f/ϵ_f'	σ_f/σ_f', (MPa/MPa)	b	c
A36 base metal	414	225/230	780/1100	/0.25	1.19/0.27	950/1015	−0.132	−0.45
A36 HAZ	667	530/400	980/1490	0.102/0.215	0.74/0.22	920/720	−0.070	−0.49
E60S weld metal	710	580/385	990/1010	0.098/0.155	0.59/0.61	990/900	−0.075	−0.55
E60 weld metal	580	410/365	850/1235	0.130/0.197	0.93/0.60	1015/1030	−0.090	−0.57
A514 base metal	938	890/600	1190/1090	0.060/0.091	0.99/0.97	1490/1305	−0.080	−0.70
A514 HAZ	1408	1180/940	2110/1765	0.092/0.103	0.75/0.78	2250/2000	−0.087	−0.71
E110S weld metal	1035	835/650	1560/2020	0.092/0.177	0.86/0.85	2210/1890	−0.115	−0.73
E110 weld metal	910	760/600	1290/1670	0.085/0.166	0.90/0.59	1660/1410	−0.079	−0.59
5083-0 aluminum base metal	294	130/290	300/580	0.129/0.114	0.36/0.40	415/710	−0.122	−0.69
5183 aluminum weld metal	299	140/270	310/510	0.133/0.072	0.40/0.58	420/640	−0.107	−0.89

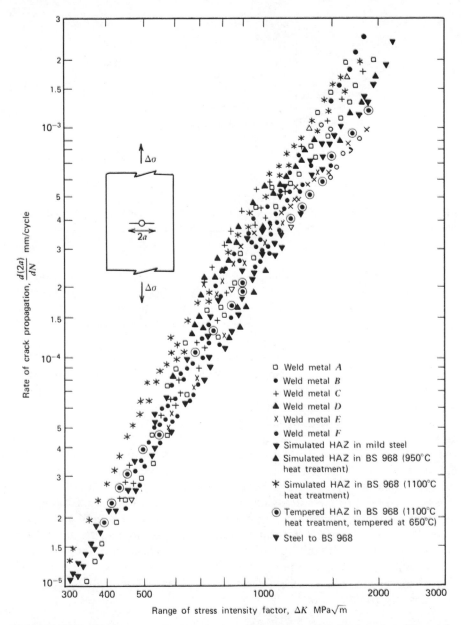

FIGURE 12.7 Fatigue crack growth data for structural C–Mn steel weld metals, HAZ and base metals [7] (reprinted with permission of the American Welding Society).

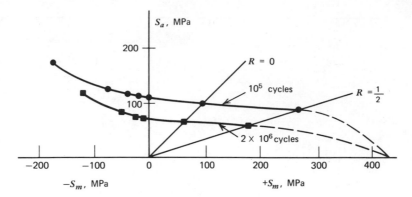

FIGURE 12.8 Constant-life diagram for carbon steel butt welds [4] (reprinted by permission of the American Society for Testing and Materials).

carbon steel butt welds is shown in Fig. 12.8, taken from Reemsnyder [4]. Tensile mean stresses do not have an appeciable affect on the allowable alternating stress and therefore it is becoming common to disregard positive S_m in weldment fatigue design recommendations. Fatigue crack growth data in weldments have also been somewhat independent of positive S_m in regions II and III. Compressive mean stresses are found to enhance fatigue resistance as shown in Fig. 12.8. The resemblance of Fig. 12.8 to the Haigh diagram for notched parts (e.g., Fig. 8.1) is quite strong. However, it is unwise to always compare compressive mean stresses in weldments to surface compressive self-stresses because of the difficulty of retaining self-stresses in the low strength weldment materials.

Spot welds usually fail in fatigue through the plate material, particularly at long lives, and through the nugget in monotonic loading. Fatigue strengths can be substantially less than those for the base material alone, and this can be attributed to high stress concentrations. Fatigue strengths generally increase with increasing size, number of spots per row, and number of rows. Grover [9] reports typical fatigue strengths ranging from 15 to 30 percent of the base metal fatigue strength for an aluminum alloy using lap joints with 1 to 3 rows of spot welds. The fatigue resistance of spot welds can be increased, however, by static overloading or by applying compression directly to each spot.

12.1.3 Improving Weldment Fatigue Resistance

We have essentially four basic ways of improving weldment fatigue resistance:

1. Improve the actual welding procedure.
2. Alter material microstructure.

3. Reduce geometrical discontinuities.
4. Induce surface compressive self-stresses.

These four methods actually overlap, since improving the welding procedure can improve the microstructure, reduce the stress concentrations, and alter self-stresses. We show earlier that substantial differences in weld metals have not provided accompanying large differences in fatigue resistance of as-welded joints. Our major emphasis thus is on reducing geometrical discontinuities and the use of self-stresses.

Reducing Geometrical Discontinuities. Removing the reinforcement bead from butt welds is the most obvious improvement. Grinding can be the quickest way to accomplish this. It adds more expense, so perhaps only the critical areas should be considered. Fillet weld toe profiles can also be improved by grinding, but since fatigue cracks often grow from either the toe or root, this may sometimes shift the failure location without increasing fatigue resistance, particularly with only partial penetration welds. Choose butt welds rather than lap joints if possible. Start/stop positions, weld ends, and even arc strikes are common locations for fatigue failures. Local grinding can improve these regions. Intersecting welds and surface ripples should be avoided. Lack of penetration essentially provides a precrack, which should be avoided. Tungsten arc inert gas dressing of welds can improve weld profile and remove entrapped inclusions at the weld toe. Large changes in stiffness should be avoided at welds. Thus good detail design can appreciably increase weldment fatigue resistance. Additional beneficial design details can be found in references 10 and 11.

Surface Compressive Self-Stresses. Common methods used for inducing surface compressive self-stresses in weldments include: shot- and hammer-peening, surface rolling, spot heating, and tensile overloading. Thermal stress relieving has been used to reduce tensile self-stresses. Each of these operations can provide a substantial increase in fatigue resistance or be ineffectual. This inconsistency is due to relaxation of the self-stresses in the lower strength weldment materials caused by repeated loading. For example, a single high compressive load can remove much of the desirable compressive self-stress. Thermal stress relieving has little influence on fatigue resistance with the condition of $R > -1$ (i.e., tensile loads > compressive loads). However, with $R < -1$ substantial improvements have been attained with stress relieving [1, 4]. Forrest [12] indicates that the fatigue strength can be improved in fillet welds by stress relieving.

It appears that mild usage of the above methods for inducing compressive self-stresses will not provide substantial improvements in fatigue resistance. Reemsnyder [4] indicated that shot-peening of non-load carrying fillet welded carbon steel and butt welded Q & T alloy steel produced 20 to 40 percent increases in S_f. However, peening to half the arc height for the above results gave only 7 percent improvement. For significant and repeatable improvements shot-peening must be closely controlled.

Weber [13] determined the effect of various methods of inducing compressive self-stresses in 7005 aluminum alloy. These results are shown in Fig. 12.9. Light shot-peening or grit blasting did not improve fatigue resistance , but a more severe hammer-peening was effective. Multiple tensile overloads were better than single overloads.

12.1.4 Weldment Fatigue Life Estimates

General procedures for estimating fatigue life are given in Chapters 8 to 10. These involve simplified procedures such as using the Haigh diagram for constant amplitude loading, as well as the more sophisticated methods involving a two stage approach and real-life load or strain histories. These same procedures can be used in weldment life estimates. Additional complications exist, however, involving the following:

1. Determining realistic values of the notch factor K_b.
2. Determining fatigue strength S_f, low cycle fatigue properties σ_f', ϵ_f', c,

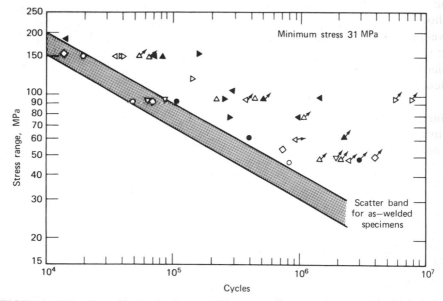

FIGURE 12.9 The effect of potential life improvement methods of non-load carrying Al–Zn–Mg (7005) fillet welds [13] (reprinted by permission of the American Society for Testing and Materials) (○) shot-peened, (▽) ground, (●) ground and shot-peened, (△) hammer-peened, (▲) ground and hammer-peened, (◁) one preload, (▷) ten preloads, (◀) one overload every 1000 cycles (▶) 10 overloads every 1000 cycles, (◇) grit blast and paint, (↗) unfailed or failure remote from transverse weld toe.

and b (Eq. 5.10), and $da/dN-\Delta K$ properties A and n (Eq. 5.17) or equivalent expressions, for weldment materials.

3. Including compressive self-stresses when relaxation may be pertinent.
4. Making basic assumptions of what size defects or cracks exist in weldments following welding.
5. Determining realistic stress intensity factors for small cracks in weldments.

Item 1 could be handled conservatively by assuming K_b which can vary from 1.0 to about 5. Thus more information on welding quality for each product is needed.

Material properties in item 2 are often similar to base metal values, which can be used as a starting point in the analysis. If actual weldment properties are known, they of course should be used. We can use data in Fig. 12.1 and Table 12.2 as reasonable design fatigue strengths of structural steels for long lives of about 2×10^6 cycles. These values can be used in connection with Haigh diagrams or modified Goodman diagrams and can be extended to shorter lives as shown in Section 8.1.2. Figure 12.1 and Table 12.2 include K_b values. The slope of the S_a versus S_m diagram for welds with positive mean stress is very shallow, as shown in Fig. 12.8 and indicated by the fatigue strength values based on S_a for $R = -1$ and 0 in Table 12.2. The magnitude of these design fatigue strengths can be extremely low. Comparison with allowable stresses based on nonpropagating cracks using S_{cat}, or ΔK_{th} from Fig. 5.23, indicates some of the recommended stress levels, though not based on S_{cat} or ΔK_{th}, appear to be consistent with threshold crack propagation ideas. A comparison of $R = 0$ fatigue strengths in Fig. 12.1 and Table 12.2 (note S_f is based on S_{max} in Fig. 12.1 and S_a in Table 12.2) for a given weldment indicates differences in the two figures of up to 35 percent. This is not unreasonable considering these design values were chosen by different societies from two different nations for very complex situations.

Self-stress relaxation (item 3) in low strength weldments is quite common. For higher strength weldment materials, however, stress relaxation is much less and hence surface compressive self-stresses can be one of the best ways to improve fatigue resistance of high strength weldment materials.

Item 4 is quite controversial, but since many defects exist in weldments, many calculations have been made based on crack growth only from an initially small defect. Yet successful estimates of fatigue life have also been made using $S-N$ or $\epsilon-N$ data. The two stage approach is quite new for weldments because of the conflict of initial defect sizes.

The primary problem of item 5 is deciding what nominal stress to use in the vicinity of a weld notch that contains a steep stress gradient. A simple and conservative approach is to use the nominal stress $\times K_t$ while the crack tip is near the notch discontinuity. This can appreciably underestimate

TABLE 12.2 Design Fatigue Strengths Based on Alternating Stress, S_a, for Steel Weldments at 2×10^6 Cycles [14].

Weld Type and Condition	A36 S_y = 248 MPa (36 ksi)		A242 S_y = 345 MPa (50 ksi)	
	$R = -1$ MPa (ksi)	$R = 0$ MPa (ksi)	$R = -1$ MPa (ksi)	$R = 0$ MPa (ksi)
1. Transverse butt weld, full penetration, as–welded	41 (6)	41 (6)	52 (7.5)	52 (7.5)
2. Transverse butt weld, full penetration, weld bead ground flush	90 (13)	76 (11)	96 (14)	86 (12.5)
3. Transverse butt weld, partial penetration, as–welded	24 (3.5)	21 (3)	——	——
4. Longitudinal butt weld, as–welded	78 (11.3)	65 (9.5)	93 (13.5)	79 (11.4)
5. Longitudinal butt weld, weld bead ground flush	90 (13)	76 (11)	117 (17)	94 (13.6)
6. Transverse fillet weld non–load carrying	41 (6)	41 (6)	41 (6)	41 (6)
7. Flexural beams reinforced with partial length cover plate fillet welds (A, B, and/or C).	28 (4)	28 (4)	28 (4)	28 (4)
8. Flexural beams with stiffeners (failure usually occurs at the stiffener weld closest to the beam tension flange)	41 (6)	41 (6)	41 (6)	41 (6)

fatigue crack growth life, however. Better K_I solutions for cracks emanating from notches, although not extensive, are given in references 7 and 15 to 17.

In summary, weldment fatigue life estimates for median lives can be made using methods from Chapters 8 to 10 and this chapter. Environmental effects on weldments can be considered from Chapter 11.

12.2 MECHANICALLY FASTENED JOINTS

12.2.1 Axially Loaded Threaded Connectors

Figure 12.10 shows a tie-rod connection and a bolted-flange connection. Both threaded members are axially loaded, but their responses to fatigue loading will be different.

The tie rod will be subject to the full variations in applied load. The bolts in the flange are preloaded to a load P. As long as the applied load L is less than P it will increase the load on the bolts only by an amount

$$\Delta L = \frac{Lb}{(b + f)} \tag{12.1}$$

where b is the stiffness of the bolt (N/m or lb/in.) and f is the stiffness of the flange. The remainder of the load L is balanced by a decrease in pressure between the flanges. The alternating stress is decreased at the expense of some increase in tensile mean stress. Whether this is beneficial or harmful depends on the ratio $b/(b + f)$ and on the mean stress influence on bolt fatigue, which is usually quite small. Reducing the bolt shank diameter and using a long bolt helps to decrease the stiffness b. The general rule is to make the bolt stiffness as low as possible and the bolted members as stiff

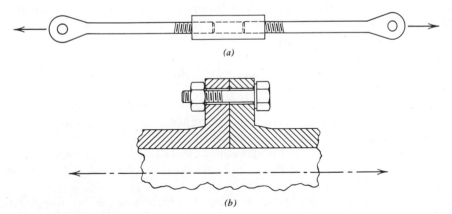

(a)

(b)

FIGURE 12.10 Axial loaded thread connectors. (*a*) Tie rod connection. (*b*) Axially loaded threaded connectors.

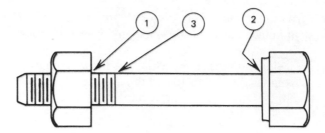

FIGURE 12.11   Fatigue critical areas in bolts.

as possible and to tighten the bolt well. One manufacturer recommends tightening bolts to achieve 70 percent of the ultimate load. Correct tightening can be verified by using special collapsing washers or by other means. Soft gaskets between the flanges would be harmful, while a member of low stiffness between the bolt head or the nut and flange would be beneficial.

Fatigue critical areas shown in Fig. 12.11 are: (*1*) the first engaged thread, (*2*) the fillet under the bolt head, and (*3*) the transition from threads to unthreaded shank. This last one is easiest to avoid by a reduced shank diameter. It may be as small as 90 percent of the thread-root diameter. This decreases the stress concentration, decreases the axial bolt stiffness b, and decreases the severity of bending stresses that can result if the bolt-head face is not exactly parallel to the flange face. The stress concentration at the first engaged thread can be alleviated by special design of the nuts to decrease the load on the first engaged thread. The theoretical stress concentration factor at that point was reported as 3.85 based on nominal diameter for coarse threads in 1943 [18] and as 9 in 1953 [19].

Special shapes can improve the fatigue strength of bolts in tension up to 40 percent. Figure 12.12, after Heywood [20], shows a connection using several design features to enhance fatigue life: (*a*) tapered nut to equalize load on engaged threads; (*b*) low stiffness collar to decrease alternating loads on the bolt; (*c*) shank reduced below thread roots to decrease stress concentration and alternating load on the bolt; (*d*) locating collar; and (*e*) tapered bolt head to decrease stress concentration and bending stresses.

Heywood [20] treats this subject thoroughly; he lists 52 references on bolts and nuts.

Rolled threads have greater fatigue resistance than cut or ground threads as is shown in Section 7.2.1. The difference is more pronounced with stronger materials. Table 12.3, from data by Heywood [20], shows typical values for alternating stress for 10 million cycles at zero to tension loading. The stresses are calculated by dividing the load by the area at the root of the threads.

The low values for cut threads suggest that these stresses are thresholds of crack propagation. Fatigue strengths of cut and rolled threads are also compared in Table 7.2, which shows much greater fatigue strengths. Several

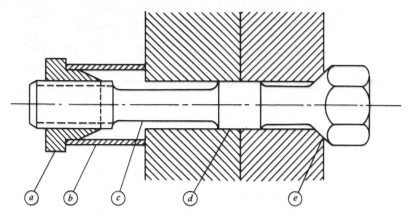

FIGURE 12.12   Means to reduce bolt fatigue failures.

TABLE 12.3   Fatigue Resistance of Cut and Rolled Threads [20]

Tensile strength—S_u, [MPa (ksi)]	500 (70)	1500 (215)
Cut thread—S_a, [MPa (ksi)]	35 (5)	50 (7)
Rolled thread—S_a, [MPa (ksi)]	35 (5)	150 (22)

items explain the differences. Table 12.3 is based on the lower bounds of scatter bands, while Table 7.2 is based on few tests. Table 12.3 is based on ordinary external bolt threads, while Table 7.2 is based on internal threads of sucker rod couplings, and Table 12.3 reports alternating stresses, while Table 7.2 reports maximum stress with a constant minimum stress of 42 MPa (6.1 ksi).

12.2.2   Pin Joints

The ends of the tie rods in Fig. 12.10a are pin joints. The stress concentrations in the loaded holes of such joints are much greater than for open holes, and they may be aggravated by fretting. Heywood [20] shows that the stress concentration factor K_t, based on the gross section of the lug (width × thickness), is at least 4.8 for the following proportions:

Pin diameter equal to d.
Width of lug equal to $2.2d$.
Height from pin center to top of lug $2.2d$ or more.

For greater or smaller width of lugs K_t becomes greater, as it does for height less than width. Heywood shows S-N curves based on the minimum cross section $(W - d)t$ of lugs with $d = 25$ mm (1 in.) and near optimum proportions. Typical lower bound fatigue strengths are shown in Table 12.4.

TABLE 12.4 Fatigue Strengths of Lugs Loaded by Pins [20]

Material	Reported Quantity	Mean Stress	Alternating Stress	
			$N = 10^4$	$N = 10^7$
Steel	S_a/S_u	$0.2S_u$	$0.12S_u$	$0.03S_u$
Aluminum	S_a MPa (ksi)	140(20)	70(10)	8 (1.2)

These very low values can be improved by decreasing the fretting action. Interference fits between pin and lugs can increase the fatigue strength threefold at 10^7 cycles. Their effect was negligible at 10^4 cycles. Clamping the joint—axially or radially—was also effective. When the pin must be free to move one can use bushings heavy enough to load the lug hole to a high compressive strain. Enlarging the hole plastically is also effective. Lubricating the pin is effective in increasing fatigue life; molybdenum disulfide paste is particularly beneficial. Decreasing fretting by changing the shape of the pin as shown in Fig. 12.13 is also effective. Frost et al. [21] quote B. C. Clarke [22] as recommending the distance between flats to be 0.7 times the pin diameter.

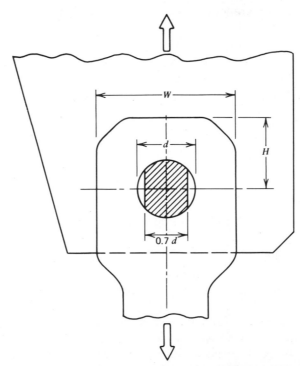

FIGURE 12.13 Lug with pin shaped for reducing fretting.

TABLE 12.5 Fatigue Strength of Bolted Connections of Plates [20]

Type[a]	Static Strength, [MPa (ksi)]	Fatigue Strength S_a, [MPa (ksi)]	Cycles
A	445 (64)	28 (4)	10^6
A	445 (64)	18 (2.6)	10^7
B	219 (31.5)	28 (4)	10^6
B	219 (31.5)	18 (2.6)	10^7
C		20 (2.8)	10^6

[a] A = double shear, two rows of two bolts; B = same as A, but scarfed (tapered) joint; C = lap joint, three rows of three bolts.

12.2.3 Riveted and Bolted Connections

The bolted or riveted joint that is strongest in a static test is not always strongest in fatigue. For well proportioned joints without special hole treatment and without special effort to transmit all the load by friction between the faying surfaces we find typical values in Table 12.5, from Heywood [20], based on stress calculated from the gross area (width × thickness of plate). The plates were aluminum, the bolts steel.

The scarfed, double-shear joint B, with a practically uniform thickness of 24.5 mm, had only half the static strength of the double-shear joint A, which was 30 mm thick, with a 15 mm tongue between the two sides, each 7.5 mm thick. However, the fatigue strengths of the two joints were equal. Both type A and type B were of aluminum alloy 75 T6 with 600 MPa (87 ksi) tensile strength. The mean stress was 49 MPa (7 ksi). The lap joint connected plates of 24 ST Alclad, S_u = 480 MPa (69 ksi), 9.5 mm thick. The bolts were 9.5 mm (0.375 in.), spaced 38 mm (1.5 in.) center to center.

Failures can be caused by fretting between bolt or rivet and hole, or by fretting of the edge of the plate, or from the stress concentration alone. Tightening the fasteners to transmit the load by friction increases the fatigue strength of the joint. Heywood [20] shows in one case more than triple the strength of "typical" aircraft joints when the torque on a 9.5 mm (0.375 in.) bolt was increased from 3.4 to 34 N.m (30 to 300 lb in.).

Radial compression of the hole also increases the fatigue strength. Shewchuk and Roberts [23] have nearly tripled the fatigue strength at $R = 0$ and 10^6 to 10^7 cycles of a double shear joint of 1.6 mm ($\frac{1}{16}$ in.) aluminum alloy by dimple-compressing the hole. The gain decreases with increasing mean stress and disappears at stress ratios (S_{min}/S_{max}), of about 0.8 as shown by Kuc and Shewchuk [24]. Speakman [25] has obtained 50 percent improvement in fatigue strength at 10^6 cycles by stress-coining the fastener holes.

12.2.4 Hub Connections

Common methods for transmitting torque between a shaft and the hub of a wheel, gear, or other member are setscrews, keys, and splines.

Setscrews have their place but should not be used to transmit alternating torques. Instead of the desirable long shank they have only a very short end between the first engaged thread and the point of loading. The force on the edge of the end of an almost loose setscrew can be calculated as $F = 2T/d$, where T is the torque and d is the diameter of the end of the setscrew (Fig. 12.14). This neglects the contribution of friction, which is unreliable. With the small contact area this invariably leads to plastic deformation. If the torque is reversed the plastic deformation on the opposite side of the contact area of the screw produces loosening. Fortunately these elements are used only for applications believed to be noncritical. But serious accidents have resulted, for instance, from loosened windshield wiper arms.

The reduction of fatigue strength produced by *keyways* and straight *splines* depends on the fillet radius. For common proportions Peterson [26] shows values of K_t between 2 and 4. We believe that K_f is not much less.

The use of *involute splines* is an excellent way to connect shafts to hubs for high fatigue strength. Torsion bars with a body diameter of 59.7 mm (2.35 in.) and a spline minor diameter of 67.9 mm (2.675 in.) failed sometimes in the body and sometimes in the splines when tested with a torque ratio R (T_{min}/T_{max}) of about 0.1 and fatigue lives of about $\frac{1}{4}$ million cycles (50,000 cycles minimum required) [27]. There were 55 splines, with 50 degree pressure angle and 1 mm (0.04 in.) root radius. Body and splines had been shot-peened, then preset. The fatigue notch factor, based on the diameter ratio 67.9:59.7, was 1.47. The hardness of these bars was about Rc 50, and that of the hub was about R_c 32.

In summary, joints made with bolts, rivets, keys, and so on have a high

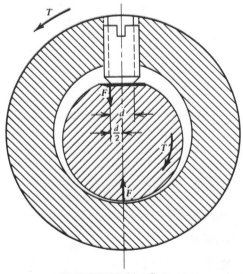

FIGURE 12.14   Setscrew.

fatigue notch factor K_f based on gross area of the joined member. The joint efficiency $1/K_f$ can be increased very substantially by careful detail design (e.g., splines with large root radii) and by prestressing (e.g., threads rolled after heat treatment).

12.3 DOS AND DON'TS IN DESIGN

12.3.1 Welds

1. Do reduce stress concentrations by grinding butt welds flush and smooth, dressing fillet welds, and avoiding undercuts and stray arc strikes.
2. Do locate joints in regions of low stress.
3. Don't use a joint with large variations in stiffness.
4. Do consider methods such as shot- or hammer-peening, surface rolling, tensile overloading, and local heating to induce desirable surface compressive self-stresses. Careful control is needed.
5. Don't use intermittent welds.
6. Do use butt welds rather than lap joints.

12.3.2 Mechanically Fastened Joints

1. Do consider preloading bolts involving flanges. A high flange stiffness and low bolt stiffness is desirable.
2. Don't neglect the details of mechanical joints since fatigue resistance can be substantially altered by small important details that may have little influence on static behavior.
3. Do use rolled threads rather than cut threads.

REFERENCES FOR CHAPTER 12

1. K. G. Richards, "Fatigue Strength of Welded Structures," The Welding Institute, Abington Hall, England May 1969.
2. British Standard, BS 153—Steel Girder Bridges.
3. P. Albrecht, "A Study of Fatigue Striations in Weld Toe Cracks," *Fatigue Testing of Weldments,* D. W. Hoeppner, Ed., ASTM STP 648, 1978, p. 197.
4. H. S. Reemsnyder, "Development and Application of Fatigue Data for Structural Steel Weldments," *Fatigue Testing of Weldments,* D. W. Hoeppner, Ed., ASTM STP 648, 1978, p. 3.
5. W. W. Sanders, Jr. and F. V. Lawrence, Jr., "Fatigue Behavior of Aluminum Alloy Weldments," *Fatigue Testing of Weldments,* D. W. Hoeppner, Ed., ASTM STP 648, 1978, p. 22.
6. Y. Higashida, J. D. Burk, and F. V. Lawrence, Jr., "Strain-Controlled Fatigue

Behavior of ASTM A36 and A514 Grade F Steels and 5083-0 Aluminum Weld Materials," *Welding J.,* Vol. 57, Nov. 1978, p. 334s.

7. S. J. Maddox, "Assessing the Significance of Flaws in Welds Subject to Fatigue," *Welding J.,* Vol. 53, Sept. 1974, p. 401s.

8. L. A. James, "Fatigue-Crack Propagation Behavior of Several Pressure Vessel Steels and Weldments," *Welding J.,* Vol. 56, Dec. 1977, p. 386s.

9. H. J. Grover, *Fatigue of Aircraft Structures,* NAVAIR 01-1A-13, 1966.

10. W. H. Munse and L. Grover, *Fatigue of Welded Structures,* Welding Research Council, New York, 1964.

11. T. R. Gurney, *Fatigue of Welded Structures,* 2nd edition, Cambridge University Press, 1979.

12. P. G. Forrest, *Fatigue of Metals,* Pergamon Press, London, 1962.

13. D. Weber, "'Evaluation of Possible Life Improvement Methods for Aluminum-Zinc-Magnesium Fillet-Welded Details," *Fatigue Testing of Weldments,* D. W. Hoeppner, Ed., ASTM STP 648, 1978, p. 73.

14. "Interim AAR Guidelines for Fatigue Analysis of Freight Cars," Association of American Railroads, Report No. R-245, May 1977.

15. F. V. Lawrence, "Estimation of Fatigue-Crack Propagation Life in Butt Welds," *Welding J.,* Vol. 52, No. 5, May 1973, p. 212s.

16. S. J. Maddox, "An analysis of Fatigue Cracks in Fillet Welded Joints," *Int. J. Fract.,* Vol. 11, No. 2, April 1975, p. 221.

17. H. Terada, "An Analysis of the Stress Intensity Factor of a Crack Perpendicular to the Welding Bead," *Eng. Fract. Mechan.,* Vol. 8, 1976, p. 441.

18. N. Hetenyi, "The Distribution of Stress in Threaded Connectors," *Proc. SESA,* Vol. 1, No. 1, p. 147.

19. A. F. C. Brown and V. M. Hickson, "A Photo-elastic Study of Stresses in Screw Threads," *Proc. IME,* Vol. 18, 1952/1953, p. 605 ff.

20. R. B. Heywood, *Designing Against Fatigue of Metals,* Reinhold, New York, 1962.

21. N. E. Frost, K. L. Marsh, and L. P. Pook, *Metal Fatigue,* Oxford University Press, London, 1974, p. 371.

22. B. C. Clarke, Ministry of Aviation, R. A. E. Tech. Report No. 66015, 1966.

23. J. Shewchuk and F. A. Roberts, "Increasing the Fatigue Strength of Loaded Holes by Dimpling," *Trans. ASME, J. Eng. Mater. Technol.,* Vol. 96, No. 3, July 1974, p. 222 ff.

24. A. P. Kuc and J. Shewchuk, "The Effect of Varying Mean Stress on the Dimpled-Loaded-Hole Fatigue Strength of 2024-T Aluminum Alloy," Paper R79-190, SESA, Spring Meeting, 1979.

25. E. R. Speakman, "Fatigue Life Improvement Through Stress Coining Methods," *Achievement of High Fatigue Resistance in Metals and Alloys,* ASTM STP 467, 1970, p. 209.

26. R. E. Peterson, *Stress Concentration Factors,* John Wiley and Sons, New York, 1974.

27. *Manual on Design and Manufacture of Torsion Bar Springs,* SAE, 1947.

PROBLEMS FOR CHAPTER 12

1. A transverse butt welded structural steel plate with width $w = 50$ mm and thickness $t = 8$ mm is subjected to an axial load $P_{min} = 3$ kN. What value of P_{max} do you recommend such that 1 million cycles can be applied without fracture for:

   (a) Full penetration as-welded.
   (b) Partial penetration as-welded.
   (c) Full penetration with weld bead excess material ground off.

2. Repeat problem 1 if only 50,000 cycles are required.

3. If the plates of problem 1 are fillet welded lap joints, and P_{min} is still 3 kN what P_{max} can be applied for (a) 1 million cycles and (b) 50,000 cycles?

4. A butt welded A36 steel plate is subjected to fully reversed bending. A small strain gage is bonded to the plate directly at the weld toe and reads a stable strain range $\Delta \epsilon = 0.02$. Using low cycle fatigue concepts, determine the expected number of cycles to the appearance of a small crack. Repeat the problem if the strain gage reads $\Delta \epsilon = 0.005$.

5. A full penetration transverse butt welded structure of steel plate with $w = 50$ mm and $t = 10$ mm was subjected to axial repeated loads. A nondestructive inspection indicated a crack existed at the weld bead toe. It was less than 1 mm deep across the section. If $P_{min} = 0$, estimate the value of P_{max} that can be applied without fracture for 50,000 additional cycles? Use a safety factor of 2.0 on life. List all your assumptions and comment on the validity of your solution.

6. What can you do to increase the life of the component in problem 5 assuming the operating load P_{max} found in problem 5 cannot be reduced?

7. If the force in the tie rod connection of Fig. 12.10a is to vary from 0 to 40 kN, 10 million times, choose materials, dimensions, and manufacturing processes for the tie rods and the connector. What safety factor did you choose?

8. Repeat problem 7 if only 20,000 cycles are to be applied.

CHAPTER 13

FATIGUE OF
MECHANICAL COMPONENTS

Springs, gears, and rolling contact bearings are produced by specialists who can know more about them than the authors or readers of this book. These parts are discussed here not to enable the reader to create a better design than the specialists, but as examples of common good practice that can serve as a guide to the design of other parts.

13.1 SPRINGS

Springs must be highly stressed so that reasonably large deflections may be obtained in reasonably small spaces. It is also desirable for economy of material, because in springs the amount of material is inversely proportional to the square of the permitted stress. Higher stresses also imply higher fundamental frequencies of vibration of the spring itself, which may be desirable to avoid resonance. Because of the high stress, springs are usually designed closer than most components to potentially dangerous stress values.

Many millions of compression coil springs made of steel are used for valve springs and suspension springs in automotive vehicles, where they are subjected to fatigue loadings. The Society of Automotive Engineers (SAE) has published recommendations for design stresses for such springs in a manual [1].

The fatigue strengths shown in the fourth edition (1973) are going to remain unchanged in the next edition (1980?).

The fatigue stress recommendations are given in the form of modified Goodman diagrams, as in Fig. 13.1. These diagrams do not show strength values but rather strength factors. The actual strength values are obtained by multiplying the ultimate tensile strength by the factor. For instance, according to Fig. 13.1 a wire of 1500 MPa ultimate tensile strength would permit torsional stress from 600 MPa initial (minimum) stress to 870 MPa maximum stress, as indicated by the dotted line, and should not be used

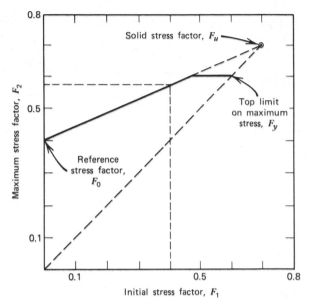

FIGURE 13.1 Fatigue stress factor diagram for helical compression springs [1] (reprinted with permission of the Society of Automotive Engineers).

with maximum stresses above 900 MPa. The stresses are to be calculated with the curvature correction (Wahl factor).

Table 13.1 gives the corner points F_u, F_y, F_0 for 15 diagrams like Fig. 13.1. A few observations are worth noting. Presetting increases the "solid" strength (factor F_u) and the "yield" strength (factor F_y) but has very little effect on the permissible stress range. F_y decreases more than 20 percent between 10^4 and 10^7 cycles life. Shot-peening increases the permissible stress range 50 percent or more for 10^7 cycles, and 10 percent or more for 10^4 cycles.

The hard drawn wire and the oil tempered wire are not likely to show the same relation of cyclic to monotonic stress–strain behavior; yet they are treated as if they were quite similar. The difference in cyclic stress–strain behavior is only one of many factors that influence fatigue life; it was disregarded in the manual.

Presetting is a process of overloading to induce favorable self-stresses. The spring is coiled longer than the specified or desired free length. It is then compressed solid several times. This decreases the free height of highly stressed springs and produces favorable self-stresses. This process is not practical for extension springs.

The ultimate tensile strengths to which the fatigue strength diagrams, Fig. 13.1, refer are much greater for smaller diameter wires than for larger diameters. Table 13.2 gives some values of ultimate tensile strength for the spring materials mentioned in Table 13.1.

TABLE 13.1   Fatigue Strength Factors for High Quality Round Wire Helical Compression Springs[a,b]

Cold Wound, Not Preset	$F_u = 0.56$		
	10^4 cycles	10^5 cycles	10^7 cycles
F_y	0.52	0.49	0.45
F_0 not peened	0.40	0.31	0.22
F_0 peened	0.44	0.40	0.34

Cold Wound, Preset	$F_u = 0.70$		
	10^4 cycles	10^5 cycles	10^7 cycles
F_y	0.65	0.62	0.56
F_0 not peened	0.41	0.33	0.24
F_0 peened	0.46	0.41	0.36

Hot Coiled, Preset, Shot-Peened	$F_u = 0.74$		
	10^4 cycles	10^5 cycles	10^7 cycles
F_0	0.50	0.44	0.36

[a] From data in reference 1.
[b] Materials for cold wound springs: Music wire; hand drawn valve spring quality wire; oil tempered valve spring quality wire. Material for hot coiled springs: steel, S_u 1180 to 2000 MPa (170 to 290 ksi).

SAE has also published a manual of leaf springs [2]. It is less explicit on permissible stresses. Among other things it shows leaf sections that save 5 to 10 percent of weight by taking advantage of the fact that for equal fatigue life the stress range on the compression side of the leaf may be higher than on the tension side. Figure 13.2 shows two sections of this type.

Torsion bars are among the most efficient springs. They are usually preset and shot-peened. The design of their ends is critical for performance in fatigue. Involute splines are often used. They are discussed briefly in Chapter 12 and more thoroughly in reference 3, which shows an efficient end connection for such springs. Flat leaves in torsion are far more efficient springs than might be expected. If the ratio of width to thickness is larger than 6 the stress distribution is quite uniform on the wide side. If the ends of such springs are held between parallel jaws they will break at the jaw because of secondary bending stresses at the transition from the twisted main length to the untwisted end. Torque should be transmitted to such springs through V-shaped slots in which the ends are inserted, as shown in Fig. 13.3.

While this is necessary to obtain good fatigue life it also improves the

TABLE 13.2   Ultimate Tensile Strengths for Wires and Bars of Some Spring Steels [MPa (ksi)][a]

| Material | Diameter, (mm): | 0.1 | 0.5 | 1 | 5 | 10 | 50 |
	(Diameter (in.)):	0.004	0.020	0.040	0.2	0.4	2
Music wire		3000	2500	2300	1750		
		(435)	(360)	(330)	(250)		
					>		
Hard drawn			2000	1800	1400	1200	
			(290)	(260)	(200)	(170)	
Hard drawn Valve spring					1600		
					(230)		
Oil tempered Valve spring			2100	1950	1600	1400	
			(300)	(280)	(230)	(200)	
Hot coiled Alloy steel						1450	1450
						(210)	(210)

[a] From data in reference 1.

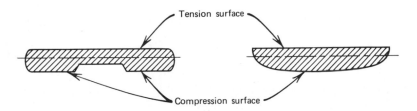

FIGURE 13.2   Special sections of leaf spring steel.

efficiency because these springs then do not have the ends, which serve only to transmit forces without contributing to the deflection of the spring. This is one more example of the importance of details in fatigue design.

13.2 ROLLING CONTACT BEARINGS

The designer of machines normally does not detail the parts of ball bearings and roller bearings that he needs but selects bearing assemblies from a catalog. The bearing catalogs list load ratings for the different bearings. These ratings are based on an expected fatigue life. It is instructive to see how they relate to our general knowledge of fatigue.

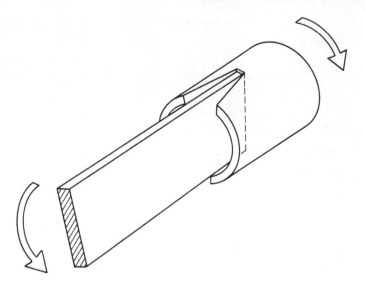

FIGURE 13.3　End fitting for a flat leaf torsion spring.

13.2.1　Life Rating Formula

One manufacturer [4], for example, shows the following

$$L = 3000 \left(\frac{C}{P}\right)^{3.33} \left(\frac{500}{S}\right) \tag{13.1}$$

where　L = life in operating hours that will be exceeded by 90 percent of
a large sample of bearings

C = catalog rating

P = load, computed with due regard for radial and thrust components

S = speed of shaft in rpm

The exponent 3.33 implies that bearing life is inversely proportional to the
tenth power of the stress, because according to the Hertz formula the stress
is proportional to the cube root of the load for balls in contact with grooves.
This corresponds closely to the exponent b, which is approximately 0.1 in
Eq. 5.11, which relates elastic strain to life. It is interesting that the same
exponent $\frac{10}{3}$ is used by the same company in their roller bearing catalog,
where it implies that life is inversely proportional to the sixth power of the
load according to the Hertz formula for cylindrical rolls. The same formula
is also used for tapered roller bearings.

In terms of cycles to failure the numbers 3000 hours and 500 rpm correspond to 90 million revolutions of the shaft and many times that number of
passages of balls over the most highly loaded point of the stationary race.

There obviously is no fatigue limit for these parts, which are made of vacuum degassed high carbon chrome steel (52100) heat treated to Rockwell C58. The life that 90 percent of a large group of bearings are expected to survive is commonly known as the B-10 life; it is used as a basis for rating by this and many other ball bearing manufacturers. The B-50 or median life is taken as five times the B-10 life according to this catalog [4]. The ball bearing industry has pioneered ratings based on explicit probabilities of survival and the use of the Weibull distribution for statistical analysis. Assuming a two parameter Weibull distribution the slope b can be calculated from Eq. 5.24 using the relation $(B$-50$) = 5(B$-10$)$ and found to be 1.17. Extrapolating to a higher reliability, say 99.9 percent survival, we would find $(B$-0.1$) = (B$-10$)/53$ by solving

$$\left(\frac{B-10}{B-0.1}\right)^{1.17} = \frac{\ln 0.9}{\ln 0.999}$$

Commenting on this, Faires [5] says that, "the actual minimum life at virtually 100 percent reliability is about 5 percent of the B-10 life" according to Harris [6]. This implies a three-parameter Weibull distribution, with a minimum life N_0 about $0.04(B$-10$)$. Equation 5.23 would then be used instead of Eq. 5.24. In addition, the roller bearing catalog [4] states that, "early failure from causes other than pure fatigue does occur sufficiently often to make predictions of $(B$-10$)$ life greater than 30,000 hours open to some question." In other words, life between 90 million and 900 million revolutions, with 50 to 90 percent reliability is predictable by Eq. 13.1, but extrapolation beyond these limits is open to questions.

13.2.2 Static Load Ratings

The load calculated from Eq. 13.1 must always be less than the load P_y that would damage the bearing by producing permanent deformation or "brinelling." The catalog [4] quotes such loads based on a mean compressive stress of 2400 MPa (350 ksi) in the most heavily loaded contact ellipse. It is interesting to note that the ratio C/P_y, where C is the load for 3000 hours at 500 rpm, is much greater for small bearings than for large bearings. For very small bearings with bore diameters between 4 and 8 mm, the ratio C/P_y is of the order of $3:4$. For very large bearings, with bore diameters between 105 and 130 mm, the ratio C/P_y is of the order of $1:3$.

Ball bearings are also subject to "false brinelling." This manifests itself in slight depressions under the balls of bearings that oscillate only a small amount, as, for instance, the bearings of turbines during shipment. The mechanism of false brinelling is related to fretting. The remedy is to rotate the turbines slowly during shipment by transforming the vibratory motion to rotation through a ratcheting device.

13.2.3   Cumulative Damage

The first rule for calculating cumulative damage was proposed for ball bearings by Palmgren [7]. The linear rule he proposed works well in this application. Both the cumulative damage rule and the statistical treatment of fatigue failures were first used in the Swedish ball bearing industry.

13.2.4   Processing Factors

The material to which the ratings in the catalog [4] apply is vacuum degassed steel. Steel produced by the consumable electrode vacuum melting process permits higher ratings [4]. This indicates the importance of microscopic inclusions as stress concentrators. Ball bearings made of less carefully cleaned steel, and with less perfect surfaces, can be obtained at less cost and may be very suitable for certain applications. Their ratings, however, will be lower if based on the same criteria.

The state of self-stress has an effect on rolling contact fatigue, as it has on other fatigue phenomena. Most investigators have reported effects similar to those in axial fatigue tests, but Kepple and Mattson [8] separated pure self-stress effects from microstructural changes and found that tensile self-stress is indeed harmful, as expected, but that in rolling contact fatigue compressive self-stress by itself is not beneficial without microstructural changes.

13.2.5   Failure Modes

Fatigue failures of ball bearings and of roller bearings manifest themselves by deterioration of the contact surfaces, which can be detected most easily by the noise created by rotation of the bearings. High alternating shear stresses of the order of 0.15 of the maximum compressive contact stress exist at a small depth below the surface. The depth at which the maximum shear stress occurs is of the order of a quarter of the width of the contact area [9]. The repeated application of these shear stresses can produce small voids below the surface: cracks grow from the voids at an angle towards the surface and produce pitting of the surface. Smaller shear stresses exist ahead and behind the contact area. Their maximum is at the surface. They change direction as the contact passes them and thus can produce a higher shear stress range than the first mentioned shear stress. The theory of rolling contact fatigue is still far from complete [10].

13.3   GEARS

The art of gear design is old and highly developed. It is based on experience generalized by analysis. There are many types of gears, and many degrees

9. S. P. Timoshenko and J. N. Goodier, *Theory of Elasticity*, McGraw-Hill Book Co., New York, 1951, pp. 372–382.

10. G. J. Moyar and JoDean Morrow, *Surface Failure of Bearings and Other Rolling Elements*, University of Illinois Engineering Experiment Station, Bulletin 468, November 1964 (bibliography lists 182 references).

11. *AGMA Standard 221.02*, Strength of Helical and Herringbone Gear Teeth, 1965.

12. "Lloyd's Register of Shipping Reduction Gearing for Propelling and Auxiliary Machinery," Appendix I of S. A. Couling, *Industrial and Marine Gearing*, John Wiley and Sons, New York, 1962.

13. *Marine Messenger*, Westinghouse Electric Corp. Marine Division, 1979, p. 16.

14. E. Buckingham, Personal communication to V. M. Faires, published in his *Design of Machine Elements*, 4th ed., The Macmillan Co., New York, 1965.

15. V. K. Sharma, G. H. Walter, and D. H. Breen, "Predicting Case Depths for Gears," *Prod. Eng.*, June 1979, p. 49.

APPENDIX A
MATERIAL PROPERTIES

TABLE A.1 Fully Reversed Rotating Bending Unnotched Fatigue Limits, S_f, of Selected Engineering Alloys

Material	Process Description	S_u, [MPa (ksi)]	S_y, [MPa (ksi)]	ϵ_f	$S_f{}^a$, [MPa (ksi)]
Steel[b]					
1015	C.D.—0 percent	455 (66)	275 (40)	1.20	240 (35)
1015	C.D.—30 percent	620 (90)	585 (85)	0.62	315 (46)
1015	C.D.—60 percent	710 (102)	605 (88)	0.78	350 (51)
1015	C.D.—80 percent	790 (115)	660 (96)	0.30	365 (53)
1020	H.R.	450 (65)	330 (48)	0.89	240 (35)
1040	H.R.	620 (90)	410 (60)	0.69	295 (43)
1040	C.D.—0 percent	670 (97)	405 (59)	0.71	345 (50)
1040	C.D.—20 percent	805 (117)	670 (97)	0.58	370 (54)
1040	C.D.—50 percent	965 (140)	855 (124)	0.33	410 (60)
1060	C.D.—stress relief	735 (107)	425 (62)	0.49	385 (56)
4340	Annealed	745 (108)	475 (69)	0.69	340 (49)
4340	Q & T (538°C)	1260 (183)	1170 (170)	0.73	670 (97)
4340	Q & T (427°C)	1530 (222)	1380 (200)	0.63	470 (68)
4340	Q & T (204°C)	1950 (283)	1640 (238)	0.65	480 (70)
HY140	Q & T (538°C)	1030 (149)	980 (142)	1.05	480 (70)
D6AC	T (260°C)	2000 (290)	1720 (250)	0.45	690 (100)
9Ni–4Co–25	T (315°C)	1930 (280)	1760 (255)	0.43	620 (90)

18 Ni marage	Age (482°C)	1540 (225)	1480 (215)	0.80	690 (100)
18 Ni marage	VM age (482°C)	1760 (256)	1630 (237)	0.97	690 (100)
18 Ni marage	VM age (482°C)	1980 (288)	1920 (279)	0.69	760 (110)
18 Ni marage	VM age (482°C)	2425 (352)	2380 (345)	0.58	760 (110)
Aluminum[c]			Based on 5×10^8 cycles		
1100-0		90 (13)	35 (5)		35 (5)
2024-T3		480 (70)	345 (50)		140 (20)
2024-T4		470 (68)	325 (47)		140 (20)
6061-T6		310 (45)	275 (40)		95 (14)
7075-T6		570 (83)	500 (73)		160 (23)
Others					
Ti[d]	Annealed	520 (75)			330 (52)
Ti-6Al-4V[e]		1190 (172)	1090 (158)		365 (53)
Copper[d]	Annealed	235 (34)	75 (11)		75 (11)
60/40 brass[d]	Annealed	315 (46)	95 (14)		85 (12)
Phosphor bronze[d]	Annealed	340 (49)	150 (22)		170 (25)

[a] Incomplete information on surface finish. These values *do not* represent design fatigue limits.
[b] *Structural Alloys Handbook*, Mechanical Properties Data Center, Traverse City, Michigan, 1977.
[c] *Aluminum Standards and Data 1976*, The Aluminum Association, New York, 1976.
[d] *Metals Handbook*, Vol. 1, ASM, 1961.
[e] *Mil-Handbook 5*, Department of Defense, Washington, D.C.

TABLE A.2A   Monotonic and Cyclic Strain Properties of Selected Engineering Alloys: SI Units[a,b]

Material	Process Description	S_u (MPa)	S_y/S_y' (MPa/MPa)	K/K' (MPa/MPa)	n/n'	ϵ_f/ϵ_f'	σ_f/σ_f' (MPa/MPa)	b	c	S_f ($2N = 10^7$), (MPa)	S_f/S_u
Steel											
1005–1009	H.R. sheet	345	262/228	531/462	0.16/0.12	1.6/0.10	848/641	−0.109	−0.39	148	0.43
1005–1009	C.D. sheet	414	400/248	524/290	0.049/0.11	1.02/0.11	841/538	−0.073	−0.41	195	0.47
1020	H.R. sheet	441	262/241	738/772	0.19/0.18	0.96/0.41	710/896	−0.12	−0.51	152	0.34
0030[c]	Cast steel	496	303/317	—/—	0.30/0.13	0.62/0.28	750/653	−0.082	−0.51	190	0.38
Man-Ten	H.R. sheet	510	393/372	—/786	0.20/0.11	1.02/0.86	814/807	−0.071	−0.65	262	0.51
1040	As forged	621	345/386	—/—	0.22/0.18	0.93/0.61	1050/1540	−0.14	−0.57	173	0.28
RQC-100	H.R. sheet	931	883/600	1172/1434	0.06/0.14	1.02/0.66	1330/1240	−0.07	−0.69	403	0.43
4142	Drawn at temp	1062	1048/745	—/—	—/0.18	0.35/0.22	1115/1450	−0.10	−0.51	310	0.28
4142	Q & T	1413	1379/827	—/—	0.051/0.17	0.66/0.45	1825/1825	−0.08	−0.75	503	0.36
4142	Q & T	1931	1724/1344	—/—	0.048/0.13	0.43/0.09	2170/2170	−0.081	−0.61	589	0.31
4340	H.R. and annealed	827	634/455	—/—	—/0.18	0.57/0.45	1090/1200	−0.095	−0.54	274	0.33
4340	Q & T	1241	1172/758	1579/—	0.066/0.14	0.84/0.73	1655/1655	−0.076	−0.62	492	0.40
4340	Q & T	1469	1372/827	—/—	—/0.15	0.48/0.48	1560/2000	−0.091	−0.60	467	0.32
9262	Annealed	924	455/524	1744/1379	0.22/0.15	0.16/0.16	1046/1046	−0.071	−0.47	348	0.38
9262	Q & T	1000	786/648	—/1358	0.14/0.12	0.41/0.41	1220/1220	−0.073	−0.60	381	0.38
Aluminum											
1100-0	As received	110	97/62	—/—	—/0.15	2.09/1.8	—/193	−0.106	−0.69	37	0.33
2024-T3	—	469	379/427	455/655	0.032/0.065	0.28/0.22	558/1100	−0.124	−0.59	151	0.32
2024-T4	—	476	303/441	807/—	0.20/0.08	0.43/0.21	634/1015	−0.11	−0.52	175	0.37
5456-H3	—	400	234/359	—/—	—/0.16	0.42/0.46	524/725	−0.11	−0.67	124	0.31
7075-T6	—	579	469/524	827/—	0.11/0.146	0.41/0.19	745/1315	−0.126	−0.52	176	0.30

[a] These values do not represent final fatigue design properties.
[b] "Technical Report on Fatigue Properties," SAE J 1099, Feb. 1975. (With permission of the Society of Automotive Engineers.)
[c] R. I. Stephens, G. Mauritzson, P. H. Benner, and D. R. Galliart, *J. Steel Casting Res.*, No. 83, July 1978.

TABLE A.2B   Monotonic and Cyclic Strain Properties of Selected Engineering Alloys: American/British units[a,b]

Material	Process Description	S_u, (ksi)	S_y/S_y' (ksi/ksi)	K/K' (ksi/ksi)	n/n'	ϵ_f/ϵ_f'	σ_f/σ_f' (ksi/ksi)	b	c	S_f $(2N=10^7)$, (ksi)	S_f/S_u
Steel											
1005–1009	H.R. sheet	50	38/33	77/67	0.16/0.12	1.6/0.10	123/93	−0.109	−0.39	21	0.43
1005–1009	C.D. sheet	60	58/36	76/71	0.049/0.11	1.02/0.11	122/78	−0.073	−0.41	28	0.47
1020	H.R. sheet	64	38/35	107/112	0.19/0.18	0.96/0.41	103/130	−0.12	−0.51	22	0.34
0030[c]	Cast steel	72	44/46	—/114	0.30/0.13	0.62/0.28	109/95	−0.082	−0.51	28	0.38
Man-Ten	H.R. sheet	74	57/54	—/—	0.20/0.11	1.02/0.86	118/117	−0.071	−0.65	38	0.51
1040	As forged	90	50/56	170/208	0.22/0.18	0.93/0.61	152/223	−0.14	−0.57	25	0.28
RQC-100	H.R. sheet	135	128/87	—/—	0.06/0.14	1.02/0.66	193/180	−0.07	−0.69	59	0.43
4142	Drawn at temp.	154	152/108	—/—	—/0.18	0.35/0.22	162/210	−0.10	−0.51	44	0.28
4142	Q & T	205	200/120	—/—	0.051/0.17	0.66/0.45	265/265	−0.08	−0.75	73	0.36
4142	Q & T	280	250/195	—/—	0.048/0.13	0.43/0.09	315/315	−0.081	−0.61	85	0.31
4340	H.R. and annealed	120	92/66	299/—	—/0.18	0.57/0.45	158/174	−0.095	−0.54	40	0.33
4340	Q & T	180	170/110	—/—	0.066/0.14	0.84/0.73	240/240	−0.076	−0.62	71	0.40
4340	Q & T	213	199/120	253/200	—/0.15	0.48/0.48	226/290	−0.091	−0.60	68	0.32
9262	Annealed	134	66/76	—/197	0.22/0.15	0.16/0.16	151/151	−0.071	−0.47	50	0.38
9262	Q & T	145	114/94	—/197	0.14/0.12	0.41/0.41	177/177	−0.073	−0.60	55	0.38
Aluminum											
1100-0	As received	16	14/9	—/—	—/0.15	2.09/1.8	—/28	−0.106	−0.69	5	0.33
2024-T3	—	68	55/62	66/95	0.032/0.065	0.28/0.22	81/160	−0.124	−0.59	22	0.32
2024-T4	—	69	44/64	117/—	0.20/0.08	0.43/0.21	92/147	−0.11	−0.52	25	0.37
5456-H3	—	58	34/52	—/—	—/0.16	0.42/0.46	76/105	−0.11	−0.67	18	0.31
7075-T6	—	84	68/76	120/—	0.11/0.146	0.41/0.19	108/191	−0.126	−0.52	25	0.30

[a] These values do not represent final fatigue design properties.

[b] "Technical Report on Fatigue Properties," SAE J 1099, Feb. 1975. (With permission of the Society of Automotive Engineers.)

[c] R. I. Stephens, G. Mauritzson, P. H. Benner, and D. R. Galliart, *J. Steel Casting Res.*, No. 83 July 1978.

TABLE A.3   Plane Strain Fracture Toughness, K_{Ic}, for Selected Engineering Alloys (Plate Stock, L-T Direction)[a,b]

Material	Process Description	S_y MPa	(ksi)	K_{Ic} MPa$\sqrt{m}$	(ksi$\sqrt{in}$)
Steel					
4340	260°C temper	1495–1640	(217–238)	50–63	(45–57)
D6AC	540°C temper	1495	(217)	102	(93)
HP 9-4-20	550°C temper	1280–1310	(186–190)	132–154	(120–140)
HP 9-4-30	540°C temper	1320–1420	(192–206)	90–115	(82–105)
10Ni(vim)	510°C temper	1770	(257)	54–56	(49–51)
18Ni (200)	Marage	1450	(210)	110	(100)
18Ni (250)	Marage	1785	(259)	88–97	(80–88)
18Ni (300)	Marage	1905	(277)	50–64	(45–58)
Aluminum					
2014-T651		435–470	(63–68)	23–27	(21–25)
2020-T651		525–540	(76–78)	22–27	(20–25)
2024-T351		370–385	(54–56)	31–44	(28–40)
2024-T851		450	(65)	23–28	(21–25)
2124-T851		440–460	(64–67)	27–36	(25–33)
2219-T851		345–360	(50–52)	36–41	(33–37)
7050-T73651		460–510	(67–74)	33–41	(30–37)
7075-T651		515–560	(75–81)	27–31	(25–28)
7075-T7351		400–455	(58–66)	31–35	(28–32)
7079-T651		525–540	(76–78)	29–33	(26–30)
7178-T651		560	(81)	26–30	(24–27)
Titanium					
T1–6A1–4V	Mill annealed	875	(127)	123	(112)
T1–6A1–4V	Recrystallized annealed	815–835	(118–121)	85–107	(77–97)

[a] *Damage Tolerant Design Handbook,* Metals and Ceramics Information Center, Battelle Labs, Columbus, OH, 1975.
[b] R. W. Hertzberg, *Deformation and Fracture Mechanics of Engineering Materials,* John Wiley and Sons, New York, 1975.

TABLE A.4 Fatigue Crack Growth Threshold, ΔK_{TH}, for Selected Engineering Alloys

Material	S_u [MPa (ksi)]	$R = K_{min}/K_{max}$	ΔK_{TH}, [MPa$\sqrt{m}$ (ksi$\sqrt{in}$)]
Mild steel[a]	430 (62)	0.13	6.6 (6.0)
		0.35	5.2 (4.7)
		0.49	4.3 (3.9)
		0.64	3.2 (2.9)
		0.75	3.8 (3.5)
A533B[b]	—	0.1	8.0 (7.3)
		0.3	5.7 (5.2)
		0.5	4.8 (4.4)
		0.7	3.1 (2.8)
		0.8	3.1 (2.8)
A508[b]	606 (88)	0.1	6.7 (6.1)
		0.5	5.6 (5.1)
		0.7	3.1 (2.8)
18/8 stainless[a]	665 (97)	0	6.0 (5.5)
		0.33	5.9 (5.4)
		0.62	4.6 (4.2)
		0.74	4.1 (3.7)
D6AC[c]	1970 (286)	0.03	3.4 (3.1)
7050-T7[c]	497 (72)	0.04	2.5 (2.3)
2219-T8[d]	—	0.1	2.7 (2.5)
		0.5	1.4 (1.3)
		0.8	1.3 (1.2)
Titanium[a]	540 (78)	0.6	2.2 (2.0)
Ti–6A1–4V[e]	1035 (150)	0.15	6.6 (6)
		0.33	4.4 (4)
Copper[a]	215 (31)	0	2.5 (2.3)
		0.33	1.8 (1.6)
		0.56	1.5 (1.4)
		0.69	1.4 (1.3)
		0.80	1.3 (1.2)
60/40 brass[b]	325 (47)	0	3.5 (3.2)
		0.33	3.1 (2.8)
		0.51	2.6 (2.4)
		0.72	2.6 (2.4)
Nickel[c]	430 (62)	0	7.9 (7.2)
		0.33	6.5 (5.9)
		0.57	5.2 (4.7)
		0.71	3.6 (3.3)

[a] N. E. Frost, K. J. Marsh, and L. P. Pook, *Metal Fatigue,* Oxford University Press, London, 1974.
[b] P. C. Paris, R. J. Bucci, E. T. Wessel, W. G. Clark, and T. R. Mager, in *Stress Analysis and Growth of Cracks,* Part I, ASTM STP 513, 1972.
[c] J. Mautz and V. Weiss in *Cracks and Fracture,* ASTM STP 601, 1976.
[d] S. J. Hudak, A. Saxena, R. J. Bucci, and R. C. Malcolm, "Development of Standard Methods of Testing and Analyzing Fatigue Crack Growth Data," Westinghouse Research Labs, March, 1977.
[e] R. J. Bucci, P. C. Paris, R. W. Hertzberg, R. A. Schmidt, and A. F. Anderson, in *Stress Analysis and Growth of Cracks,* Part I, ASTM STP 513, 1972.

TABLE A.5 Corrosion Fatigue Behavior in Water or Salt Water for Life Greater than 10^7 cycles for Selected Engineering Alloys^a

Material	Process Description	S_u, [MPa (ksi)]	Frequency, (Hz)	Corrosive Media	Fatigue Limit in Air, [MPa (ksi)]	Corrosion Fatigue Strength, [MPa (ksi)]	Corrosion Fatigue Strength ÷ Air Fatigue Limit
Mild Steel	Normalized			River water	260 (38)	32 (4.6)	0.12
0.21% C steel	Annealed	490 (71)	22	Sea H$_2$O	220 (32)	30 (4.3)	0.13
1035		600 (87)	29	6.8 percent salt water; complete immersion	280 (41)	170 (25)	0.61
1050		650 (94)			220 (32)	140 (20)	0.63
1050	Q & T	900 (130)			415 (60)	170 (25)	0.42
4130	Q & T	880 (128)			480 (70)	190 (27)	0.38
9260	Normalized	985 (143)			500 (72)	170 (25)	0.35
5Cr steel	Q & T	990 (130)			510 (74)	365 (53)	0.71
Wrought iron		325 (47)			210 (30)	130 (19)	0.64
12.5Cr steel	Annealed	1000 (146)	6	Fresh water in torsion	225 (37)	125 (18)	0.49
18/8 stainless	Annealed	1300 (188)			195 (28)	85 (12)	0.44
18.5Cr steel	Annealed	770 (112)			240 (35)	195 (28)	0.79

Material	Condition			Environment			
Aluminum	Annealed	90 (13)	24	River water with saline solution about one-half of sea water	19 (2.70)	7.6 (1.1)	0.41
Copper	Annealed	215 (31)			69 (10)	70 (10.5)	1.04
Monel	Annealed	565 (82)			250 (36)	200 (29)	0.81
60/40 brass	Annealed	365 (53)			150 (22)	130 (19)	0.85
Nickel	Annealed	530 (77)			235 (34)	160 (23)	0.68
Phosphor bronze	Normalized	430 (62)	37	3 percent salt spray	150 (22)	180 (26)	1.20
Aluminum–3Mg	Heat treated	—	83	3 percent salt spray	125 (18)	48 (7)	0.38
Aluminum–7Mg	Heat treated	—	83	3 percent salt spray	110 (16)	48 (7)	0.45
Aluminum–Cu Mg	Heat treated	—	83	3 percent salt spray	180 (26)	85 (12)	0.47
Mg–Al–Zn AZG		—		Tap H_2O	75 (11)	40 (6)	0.49
Mg–Al–Mn AZ537		—		Tap H_2O	75 (11)	55 (8)	0.68
Mg–Al–Mn AM503		—		3 percent salt H_2O	55 (8)	20 (3)	0.36
Mg–Al–Mn AZM		—		3 percent salt H_2O	150 (22)	14 (2)	0.08

[a] Reprinted from P. G. Forrest, *Fatigue of Metals*, Pergamon Press, 1962 (with permission of Pergamon Press.) All test specimens rotating bending unless specified. These results have been collected from various sources and only illustrate corrosion fatigue effects and are not to be used as design values.

APPENDIX B

QUANTITATIVE DATA ON SCATTER IN FATIGUE OF METALS

The data are given in terms of C, the coefficient of variation of the fatigue strength and of D, the standard deviation of the logarithm of the fatigue life. D is defined as

$$D = \frac{\sum(L_i - \bar{L})^{1/2}}{n+1} \quad \left(\frac{\sum(L_i - \bar{L})^2}{n-1}\right)^{1/2}$$

where n = number of specimens

$\bar{L} = \sum L_i / n$

$L_i = \log_{10} N_i$

N_i = fatigue life

D is a logarithm and denotes the ratio of the logarithmic mean fatigue life to the fatigue life that 16 percent of all specimens can be expected to survive. For instance, $D = 0.5$ means a ratio of 3, or 16 percent of all specimens can be expected to survive one-third of the logarithmic mean life.

The coefficient of variation C is defined as

$$C = \frac{\text{SD}}{M}$$

where SD = the standard deviation of the fatigue strength

M = the mean fatigue strength

These measures of scatter would permit estimates of the life for which we must expect X percent failure at a given stress range or of the stress for which X percent would fail before a given life, if we knew the distribution function. For example, for a normal distribution of fatigue strength and a log-normal distribution of fatigue lives we can compute for 10 percent expected to fail:

$$\text{stress} = (1 - 1.3C) \text{ mean strength}$$
$$\log \text{life} = (\log \text{ median life}) - 1.3D$$

where the factor 1.3 is taken from a table of the normal (cumulative) distribution function as the deviate corresponding to .1 probability.

We might mention here that the usual assumptions of normal distribution of strength and log-normal distribution of lives are mathematically compatible only with certain restrictions on the relation of fatigue life to stress.

A typical value $D = 0.25$ means that the logarithm of the B-10 life, for which we expect 10 percent failures, is $1.3 \times 0.25 = 0.325$ smaller than the logarithm of the median life. The B-10 life then is $10^{-0.325} = 0.47 = 47$ percent of the median life.

The coefficient of variation C is the ratio of standard deviation to mean. If life N and strength S (in terms of stress or strain) are related by a power function $N = AS^{-m}$ with constant m for all probabilities of failure, we can compute:

$$C = 10^{D/m} - 1$$

For typical values such as $D = 0.25$ and $m = 6$ we would find $C = 10$ percent, which is also fairly typical. To obtain 10 percent probability of failure we would use a stress that is 1.3×10 percent $= 13$ percent lower than the mean value.

Although D and C are good measures of the scatter between 10 and 50 percent failures they cannot be used to extrapolate directly to 0.1 percent or fewer failures. We have no reason to assume that the tail of the distribution is Gaussian or Weibullian, and we do have some reason to believe that to achieve 0.1 percent failures we must decrease the stress or the life more than predicted by log-normal or Weibullian distributions [1].

The data presented in this appendix were not derived by rigorous statistical analysis, but they are believed to be reasonably close to rigorously derived values. They show that the scatter of fatigue data defies generalized rules. Typical round values are $D = 0.25$ and $C = 0.1$, but values of D vary from 0.08 to 0.8 and values of C vary from 0.01 to 0.25.

Values of D in Table B.2 are for lives of less than 10^6 cycles because lognormal distribution with constant D does not fit the data for long lives. Schuette [2] has suggested that a normal distribution of $\log(1/N)$ would avoid the difficulties to which the assumption of normal distribution of log N leads. The probability distribution of fatigue strength at constant life is equally well fitted at long life and at short life by the assumption of normal distribution.

Table B.1 gives coefficients of variation for steel, taken from several sources, arranged in order of increasing tensile strength. They range from 3 to more than 13 percent, with no obvious trend.

TABLE B.1 Coefficients of Variation of the Fatigue Strength of Steel (percent)

Type (N = Notched)	S_u [MPa (ksi)]	Log Life (E = fatigue limit)	C, percent Source	
Armco Iron	290 (42)	E	7.8	4
950	500 (73)	5	10.6	5
950	500 (73)	7	13.9	5
Sheet steel	520 (75)	5	1.9	6
1050	630 (91)	7	2.5	4
4340	720 (103)	5	2.4	4
4340	720 (103)	5	5	4
4340	720 (103)	E	4	4
4340	870 (125)	E	3	4
4340	870 (125)	E	11.3	4
4340	958 (138)	7	6.3	7
4340 N	999 (144)	7	5	8
4340	999 (144)	7	6.4	8
4340 coated	1110 (160)	5	2.4	9
4340	1110 (160)	5	7.7	9
4340 N	1110 (160)	5	11.9	9
4340 N and N coated	1110 (160)	5	3	9
4340	1150 (167)	5	4	4
4340	1150 (167)	E	7.3	4
4340	1305 (188)	7	5.5	10
4340 N	1305 (188)	11	7	10
4340	1320 (190)	7	5.4	7
4340	1332 (192)	7	7.9	8
4340 N	1332 (192)	7	4.9	8
4340	1375 (198)	E	6.1	11
4340	1390 (200)	E	8.4	9
4340	1400 (202)	E	8.8	11
Sheet steel	1480 (211)	5	2.4	7
4340	1530 (220)	E	5.1	11
4340	1600 (230)	E	9.7	11
4340	1650 (238)	7	6.4	10
4340 N	1650 (238)	7	4	10
4340	1800 (260)	5	5.7	7
4340 N	1800 (260)	5	5.3	7
4340	1800 (260)	7	6.4	7
4340 N	1800 (260)	7	13.3	7
4340	1860 (268)	7	8.9	8
4340 N	1860 (268)	7	13.2	8
4340	2070 (298)	E	4.3	10

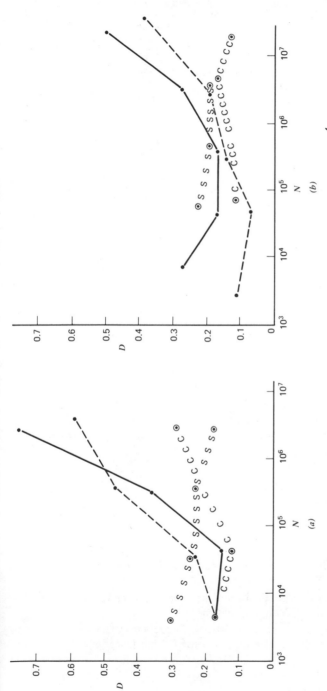

FIGURE B.1. Logarithmic standard deviation D versus cycles to failure N for aluminum alloys.
(a) Uniform cycles, (b) spectrum loading. Each point is derived from n specimens in K groups:

	K	n
Min	4	17
Max	125	936

(———•) unnotched specimens; (•– – – –•) notched specimens; (OCCCCCO) structural
components; (OSSSSO) full scale structures. (From P. R. Abelkis *Fatigue strength design and
analysis of aircraft structures*, AFFDL-TR 66-197 June 1967.)

TABLE B.2 Logarithmic Standard Deviation *D* of Fatigue Lives from [3] and Coefficients of Variation *C* of Fatigue Strengths Derived from Curves in [3]

Material	Notched		Un-notched	
	D	*C-%*	*D*	*C-%*
Aluminum alloys				
2024–T3 sheets	0.27	7	0.21	9
2024–T4 bars & rods	0.306	10	0.166	5
7075–T6 sheets	0.197	6	0.207	8
7075–T6 clad sheets			0.124	6.5
7075–T6 bars	0.30	12	0.383	19
Steel				
Billets and forgings.				
300M	0.134	10	0.143	14
Titanium alloys				
Sheet, castings, plates.				
Ti-6A1-4V-STA	.56	18	0.30	23

Table B.2 gives coefficients of variation for some aircraft materials, notched and unnotched. They are derived from consolidated data by Rice et al. [3].

Figure B.1 shows the logarithmic standard deviations of fatigue lives from pooled data by Abelkis [1]. They are plotted versus fatigue lives. Data from uniform cycling tests and from spectrum loading tests are shown separately. Structural components and full scale structures appear to have less scatter at long lives than laboratory specimens. A value of 0.2 seems reasonably typical but not conservative.

REFERENCES FOR APPENDIX B

I. Text References

1. P. R. Abelkis, "Fatigue Strength Design and Analysis of Aircraft Structures," AFFDL-TR 66-197, June 1967.
2. E. H. Schuette, "A Simplified Statistical Procedure for Obtaining Design Level Fatigue Curves," *Proc. ASTM,* Vol. 54, 1954, p. 853.
3. R. C. Rice, K, B. Davies, C. E. Jaske, and C. E. Feddersen, "Consolidation of Fatigue and Fatigue-Crack-Propagation Data for Design Use," NASA CR 2586, Oct. 1975.

II. References for Table B.1

4. E. Epremian and R. F. Mehl, "Investigation of Statistical Nature of Fatigue Properties" NACA TN 2719, June 1952.

5. R. Todd, Private communication based on data by J. M. Holt, U.S. Steel Project No. 57.12.903 and 57.019.903, 1963/1964.

6. W. E. Hering and C. W. Gadd, "Experimental Study of Fatigue Life Variability," GM Research Publication GMR 555, May 1966.

7. H. N. Cummings, "Some Quantitative Aspects of Fatigue of Metals," WADD Tech. Report 60-42, July 1960 (Dept. of Commerce PB 171084), contains data from many sources such as H. N. Cummings, F. B. Stulen, and W. C. Schulte, "Research on Ferrous Metal Fatigue," WADC-TR 58-43, Aug. 1958.

8. H. N. Cummings, F. B. Stulen, W. C. Schulte, "Investigation of Materials Fatigue Problems Applicable to Propeller Design," WADC TR 54-531, May 1955 and Oct. 1955.

9. W. L. Holshouser and H. P. Uteck, "Effect of Oleophobic Films on Fatigue Crack Propagation," ASTM preprint, 1961.

10. H. N. Cummings, F. B. Stulen, W. C. Schulte, "Investigation of Materials Fatigue Problems," WADC TR 56-611, March 1957.

11. W. L. Starkey, S. M. Marco, and R. R. Gatts, "Statistical Evaluation of Variation in Endurance Limit Among Several Heats of Propeller-Type Steel," WADC TR 55-483 ASTIA AD97190, Aug. 1956.

AUTHOR INDEX

SUBJECT INDEX